ANCIENT BOTANY

Gavin Hardy and Laurence Totelin have brought together their botanical and historical knowledge to produce this unique overview of ancient botany. It examines all the founding texts of botanical science, such as Theophrastus' *Enquiry into Plants*, Dioscorides' *Materia Medica*, Pliny the Elder's *Natural History*, Nicolaus of Damascus' *On Plants*, and Galen's *On Simple Remedies*, but also includes lesser-known texts ranging from the sixth century BCE to the seventh century CE, as well as some material evidence. The authors adopt a thematic approach rather than a chronological one, considering important issues such as the definition of a plant, nomenclature, classification, physiology, the link between plants and their environment, and the numerous usages of plants in the ancient world. The book also takes care to place ancient botany in its historical, social and economic context. The authors have explained all technical botanical terms and ancient history notions, and as a result, this work will appeal to historians of ancient science, medicine and technology; classicists; and botanists interested in the history of their discipline.

Gavin Hardy is a Tutor in the Office of Lifelong Learning at the University of Edinburgh, UK.

Laurence Totelin is Senior Lecturer in Ancient History at Cardiff University, UK.

SCIENCES OF ANTIQUITY
Series editor: Liba Taub
Director, The Whipple Museum of the History of Science, University of Cambridge

Sciences of Antiquity is a series designed to cover the subject matter of what we call science. The volumes discuss how the ancients saw, interpreted and handled the natural world, from the elements to the most complex of living things. Their discussions on these matters formed a resource for those who later worked on the same topics, including scientists. The intention of this series is to show what it was in the aims, expectations, problems and circumstances of the ancient writers that formed the nature of what they wrote. A consequent purpose is to provide historians with an understanding of the materials out of which later writers, rather than passively receiving and transmitting ancient 'ideas', constructed their own world-view.

MEDICINE IN ANTIQUITY
Vivian Nutton

TIME IN ANTIQUITY
Robert Hannah

ANCIENT ASTROLOGY
Tamsyn Barton

ANCIENT NATURAL HISTORY
Roger French

COSMOLOGY IN ANTIQUITY
M. R. Wright

ANCIENT MATHEMATICS
S. Cuomo

ANCIENT METEOROLOGY
Liba Taub

ANCIENT BOTANY

Gavin Hardy and Laurence Totelin

Routledge
Taylor & Francis Group
LONDON AND NEW YORK

First published 2016
by Routledge
2 Park Square, Milton Park, Abingdon, Oxon OX14 4RN

and by Routledge
711 Third Avenue, New York, NY 10017

Routledge is an imprint of the Taylor & Francis Group, an informa business

British Library Cataloguing-in-Publication Data
A catalogue record for this book is available from the British Library

Library of Congress Cataloging-in-Publication Data
Hardy, F. G.
Ancient botany / Gavin Hardy and Laurence M.V. Totelin.
pages cm. – (Sciences of antiquity)
Includes bibliographical references and indexes.
1. Botany, Medical–Greece–History. 2. Botany, Medical–Rome–History. 3. Botany–Greece–History. 4. Botany–Rome–History. 5. Materia medica, Vegetable. 6. Medicine, Greek and Roman. I. Totelin, Laurence M. V. II. Title.
RS63.H37 2015
615.109456'32–dc23 2015012293

ISBN 978-0-415-31119-9 (hbk)
ISBN 978-0-415-31120-5 (pbk)
ISBN 978-0-203-45835-8 (ebk)

Typeset in Times New Roman
by Wearset Ltd, Boldon, Tyne and Wear

CONTENTS

ILLUSTRATIONS

Maps

Figures

ACKNOWLEDGEMENTS

This book was a long time in the making. In 2000, Roger French and Gavin Hardy planned to write a book on *Ancient Botany* for the Routledge Series on Ancient Science. Unfortunately, Roger French passed away shortly after work was initiated on the project, on May 14, 2002. Gavin Hardy produced a first draft of the book, but felt the project would benefit from the skills of a historian of ancient science. In 2010, Laurence Totelin came on board and added her classical expertise to the project. She gave the book its current chapter organisation and integrated Gavin's initial draft into the book you are now reading.

Many people contributed to the writing of this book. Our heartfelt thanks go to Liba Taub, who put us in touch and offered detailed comments on the entire manuscript. Philip van der Eijk, Michael Herchenbach, James Longrigg, Dunstan Lowe, Elizabeth Macaulay-Lewis, Vivian Nutton, Thomas Rütten and John Scarborough answered numerous questions and provided information on various points of detail. Caroline Musgrove read through the entire manuscript, smoothing out style and checking references. We are extremely thankful for her help. Philip Hughes assisted with the drawing of the maps.

Laurence presented papers that became chapters of this book to audiences at Newcastle, School of History, Classics and Archaeology; at Cambridge, Department of History and Philosophy of Science; at the Annual Colloquium of Classics and Ancient History at Gregynog Hall, Wales; at the Ancient Greek and Roman Scientific, Medical and Technical Writing workshops, organised by the Excellence Cluster Topoi, Berlin; and at Princeton, Department of Classics. We wish to thank these audiences for their perceptive questions.

Our thanks also go to our students: students attending Gavin's courses on 'Ancient Botany' and 'The Development of Medical Botany' in the Office of Lifelong Learning at the University of Edinburgh; and students attending Laurence's course on 'Greek and Roman Medicine' and 'Science and Technology in the Graeco-Roman World' in the School of History, Archaeology and Religion, Cardiff University.

We benefited immensely from the help of the staff at the numerous libraries where we conducted our research, and at the libraries and museums where we ordered the images that illustrate this book. Particular thanks go to the staff at the Wellcome Library, London and at the Österreichische Nationalbibliothek, Vienna.

On a more personal note, our thanks go to our colleagues who discussed our ideas on a more informal basis. In particular, Gavin wishes to thank Martin Wheeler, and Laurence wishes to thank Guy Bradley, Kate Gilliver, Fay Glinister, Stephen Lambert, Shaun Tougher, Ruth Westgate and James Whitley. Laurence is also grateful to Thomas and Gwilym for their patience.

Finally, when writing this book, we terribly missed the input of two scholars: Roger French and Robert (Bob) Sharples. This book is dedicated to their memory.

NOTE TO THE READER

This book will, we hope, appeal both to students of ancient Greek and Roman societies and to modern botanists with an interest in the history of their discipline. Conventions that are familiar to one group of readers may not be to the other, and for this reason we have chosen to spell things out here.

We have limited our use of endnotes to a minimum, opting instead for in-text references. We have used endnotes when we felt in-text references would hinder the flow of reading, or to add suggestions for further reading.

We use the abbreviations BCE and CE to refer, respectively, to 'before the Common Era' and 'of the Common Era'. When obvious, we have omitted 'CE'.

A large number of ancient authors are mentioned in this book. Whenever possible, we have added dates and some biographical information. The reader will find more information on these ancient authors in *The Encyclopedia of Ancient Natural Scientists: The Greek Tradition and its Many Heirs*, edited by Paul T. Keyser and Georgia L. Irby-Massie (2008), to which we refer extensively.

Classical studies conventions

The texts discussed in this book were written, for the most part, in ancient Greek and in Latin. Unless stated otherwise, all translations from the Greek and Latin were produced by Laurence Totelin. Within translations, square brackets are used to supplement words in English that are not found in the Greek or Latin text. Important Greek and Latin words are given in round brackets. Both Greek and Latin are languages with declensions. This means that the same noun or adjective has different endings according to its case (its grammatical function). When we give a Greek or Latin word within brackets in our translation, we give it in its original grammatical form.

We transliterate Greek terms, that is, we render them in the Latin alphabet. There are various systems of transliterations, all imperfect. Here, the letters eta and omega are rendered respectively as 'ē' and 'ō'. The letter upsilon is rendered as 'u'. The letters kappa and chi are rendered as 'k' and 'ch'. The rough breathing at the beginning of words is rendered by the letter 'h'. Accents are not rendered, even though ancient Greek is an accentuated language. All Greek personal names are Latinised. That is, we chose to talk about 'Theophrastus' rather than 'Theophrastos'.

We refer to ancient texts using an English title, and limiting our use of abbreviations to the strictest minimum, as this is probably the most confusing aspect of classical studies to the non-specialist. However, since it is customary among classicists to refer to both ancient Greek and Latin texts using Latin titles, we have mentioned these in bracket at the first occurrence of a text. For a handful of titles, we use a transliterated form of the Greek title: *Alexipharmaka*; *Theriaka*; and *Geoponika*.

It is conventional to refer to an ancient text by giving a book, chapter, and in some cases, paragraph within a chapter, as in the following example: Theophrastus, *Enquiry into Plants* 2.1.3, where '2' refers to book two, '1' to chapter one, and '3' to the paragraph. For other texts, additional conventions are used. Thus, it is conventional to refer to a passage in the works of Aristotle by giving a book number, a paragraph number and the 'Bekker number' (in reference to the authoritative nineteenth-century edition of Immanuel Bekker), which consist in a number (a page number), the letter 'a' or 'b' (referring to a column), and a number between 1 and 35 (referring to a line number). For instance, Aristotle, *Generation of Animals* 1.18, 726a7, where '1' refers to the book number; '18' to the chapter number; '726' to the page in Bekker's edition; 'a' to column a in Bekker's edition; and '7' to the line in Bekker's edition. Bekker numbers are printed in the margins of all modern editions of Aristotle. In the case of texts by Plato and Plutarch, it is conventional to give the 'Stephanus number', that is, a page number and page section (a–d) in the sixteenth-century edition by Henri Estienne. For instance, we refer to Plato, *Timaeus* 77a, where '77' is the page in Etienne's edition, and 'a' is the section of that page in Etienne's edition.

In the case of Greek and Latin medical texts, which are famously difficult to navigate, it is conventional to give a reference to one or several modern editions. This can be done in a number of ways. Here we have opted for the following. In the case of works from the Hippocratic Corpus, we have given a reference to the edition in the Loeb Classical Library (when available) and to the nineteenth-century edition by Émile Littré (1839–1861). For instance, in the reference 'Hippocratic Corpus, *Nature of the Child* 22, Loeb 10.60 Potter = 7.514 Littré', 'Loeb 10' refers to volume 10 in the edition of the Hippocratic Corpus in the Loeb Classical Library; '60' refers to the page in that edition; and 'Potter' refers to 'Paul Potter', the editor of that text; '7' refers to the volume in Littré's edition; and '514' refers to the page in Littré's edition. In the case of works attributed to Galen, we have given a reference to the nineteenth-century edition by Karl Gottlob Kühn (1821–1833). For instance, in the reference 'Galen, *Properties of Foodstuffs* 1.37, 6.552 Kühn, '6' refers to the volume in Kühn's edition, and '552' to the page in Kühn's edition. For the works of Oribasius, Aetius and Paul of Aegina, we have given references to the *Corpus Medicorum Graecorum* editions. For instance, in the reference Oribasius, *Medical Collection* 11.1.1, CMG 6.1.2, p. 80 Raeder, 'CMG 6.1.2' refers to the number given to this edition in the

Corpus Medicorum Graecorum, 'Raeder' is the name of the editor of this text, and 'p. 80' is the page number in this volume of the CMG.

As the number of papyri mentioned in this book is relatively limited, we have referred to these in abbreviated form, as is conventional among classicists. All abbreviations are given in the list of abbreviations on p. xiv. When available, we have also given the Trismegistos number of the papyrus. Trismegistos is an online catalogue of papyri and inscriptions (www.trismegistos.org, accessed March 2015). In the case of literary papyri, and when available, we have also given the CeDoPal number of the papyrus. CeDoPal is an online catalogue of literary papyrus developed at the Université de Liège (Belgium) (http://web.philo.ulg.ac.be/cedopal, accessed March 2015).

Botanical conventions

Identifications of plants named in ancient text are a difficult issue (see Chapter 4), and in this book, we do not offer new identifications. When we mention possible identifications, we do so by using the conventions outlined in the International Plant Names Index, which is available online (www.ipni.org, accessed March 2015). For instance, we will talk about *Helleborus cyclophyllus* Boiss. (family: Ranunculaceae) – our hellebore – where '*Helleborus*' is the name of the genus to which this plant belongs; '*cyclophyllus*' is the name of this plant's species; and 'Boiss.' is the abbreviation of 'Boissier', referring to Pierre Edmond Boissier, a nineteenth-century Swiss botanist who worked extensively on plant taxonomy. We have made extensive use of Mabberley's *Plant Book* (third edition, 2008) for general information on plant names and classification.

ABBREVIATIONS

CIL	*Corpus inscriptionum Latinarum*, an edition of Latin inscriptions
CMG	*Corpus Medicorum Graecorum*, a series of editions of Greek medical texts formerly edited by Teubner (Berlin and Leipzig: 1908–1944) and currently edited by Akademie Verlag (Berlin).
CML	*Corpus Medicorum Latinorum*, a series of editions of Latin medical texts formerly edited by Teubner (Berlin and Leipzig: 1908–1944) and currently edited by Akademie Verlag (Berlin).
CP	refers to Theophrastus' *Causes of Plant Phaenomena*, whose Latin title is *De causis plantarum*
DK	refers to the edition of the fragments of the pre-Socratic philosophers by Hermann Diels and Walther Kranz (1952)
FgrH	refers to the edition of fragments of Greek historians by Felix Jacoby (1923–1958)
GA	refers to Aristotle's *Generation of Animals*, whose Latin title *De generatione animalium*
HN	refers to Pliny the Elder's *Natural History*, whose Latin title is *Historia Naturalis*
HP	refers to Theophrastus' *Enquiry into Plants*, whose Latin title is *Historia plantarum*
MM	refers to Dioscorides' *Materia Medica*, whose Latin title is *De material medica*
OA	refers to Palladius' *Work of Agriculture*, whose Latin title is *Opus agriculturae*
P.Ant.	papyrus from Antinoopolis
P.Cairo Zen.	papyrus kept in Cairo and belonging to the Zeno archive
P.Oxy.	papyrus from Oxyrhynchus
P.Tebt.	papyrus from Tebtunis.
PDM	*Papyri Demoticae Magicae*, the Demotic Magical Papyri
PGM	*Papyri Graecae Magicae*, the Greek Magical Papyri
RR	refers to Cato's, Varro's and Columella's works *On Agriculture*, whose Latin titles are *Res rusticae* (or *De re rustica*)

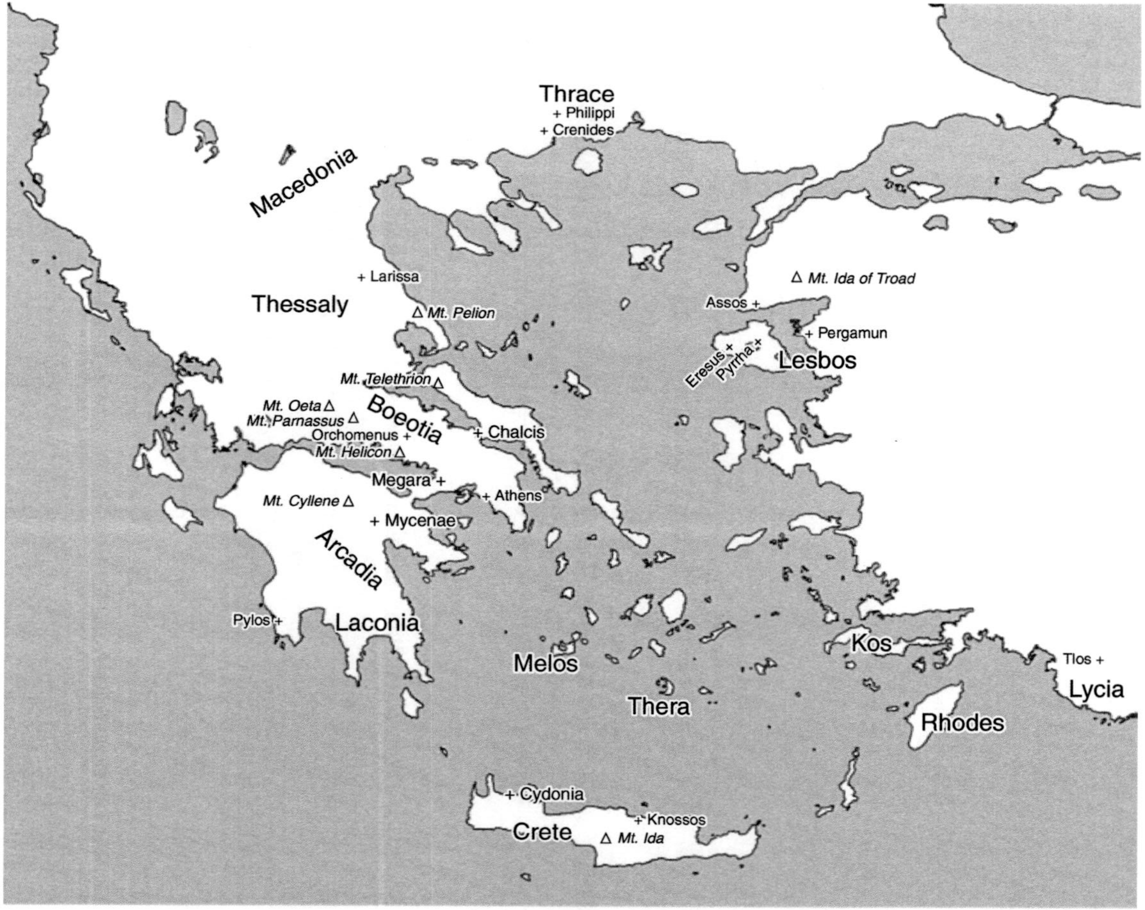

Map 1 The Greek world (source: map by Lencer via Wikimedia).

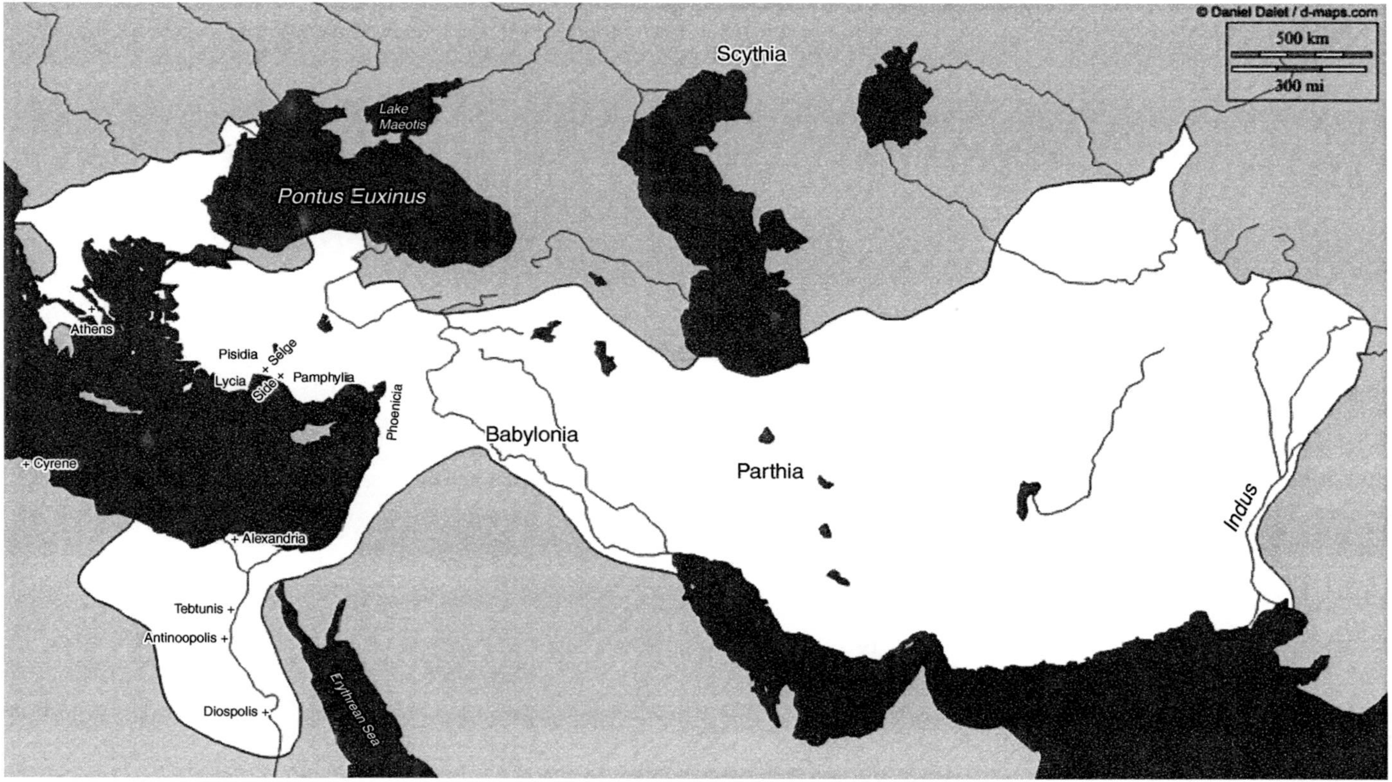

Map 2 The conquests of Alexander (source: map by D-Maps (http://d-maps.com/carte.php?num_car=1968&lang=en)).

Map 3 The Roman Empire (source: map by ColdEel via Wikimedia).

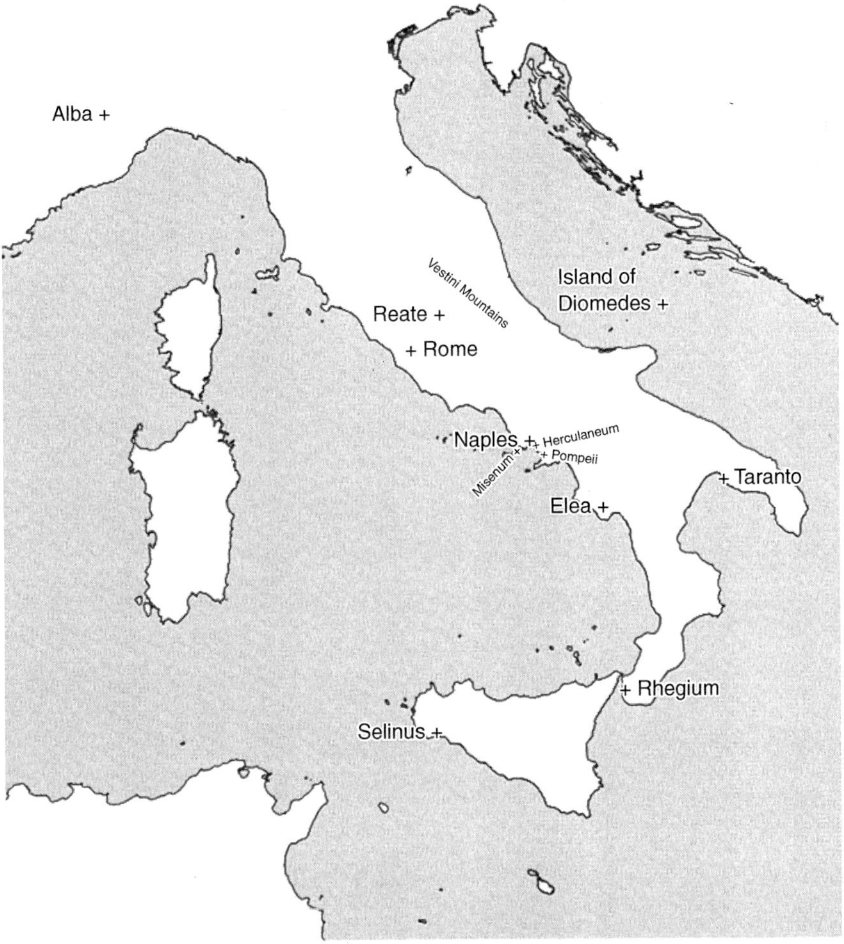

Map 4 Close-up of Italy. This map is a close-up of the area indicated by the square in Map 3 (source: map by TUBS via Wikimedia).

1

INTRODUCTION

> Even though many authors, both ancient and modern, have composed works about the preparation, powers, and testing of drugs, dear Arius, I shall attempt to show you that it is not a vain and irrational impulse that took hold of me in dealing with this topic. I am doing so because some did not give complete accounts, while others based their writings mostly on [written] enquiries (*historias*). For instance Iollas of Bithynia and Heracleides of Tarentum touched upon only a small part of the subject, leaving aside completely **the botanical tradition** (*tēn botanikēn paradosin*), and they did not mention minerals and spices at all.
>
> (Dioscorides, *Materia Medica*, preface 1, our emphasis)

1.1 General introduction

We open this book with a passage from Dioscorides, the first-century CE pharmacologist, for several reasons. First, it is one of the rare ancient texts that mentions a botanical tradition (*botanikē paradosis*), sometimes also referred to as *botanikē technē*, the botanical art.[1] Dioscorides is here criticising the pharmacologists Iollas (second century BCE, see Jacques 2008b for references) and Heracleides (first century BCE, see Stok 2008b for references) for writing drug recipes without studying that art, without examining plants in any detail. While we conceive of 'botany' as a science that deals with all plants, the ancient *botanikē technē* was the study and application of medicinally active plants. In practice, that covered most plants, since the Greeks and Romans made use of most vegetables for pharmacological purposes. This close association between plant science and medicine perpetuated through the centuries. The Renaissance witnessed the flourishing of 'physic gardens', the predecessors of our botanical gardens; and the first chairs of botany in universities were called chairs of *materia medica*. The science of botany thus had a very pragmatic beginning: it grew out of the knowledge of herbs for medical purposes. In fact, all ancient writings that deal with plants insist on their utility. In this context, drawing a distinction between 'pure' botany (i.e. the study of plants for their own sake) and 'applied' botany (i.e. the study of plants for practical purposes) is almost meaningless for the ancient world. In the present book, 'botany' refers to all technical knowledge of plants, whether it has practical applications or not.

Distinguishing between 'pure' and 'applied' ancient botany, however, has a long history. In 1694, Joseph Pitton de Tournefort (1656–1708) opened his seminal work *Elemens de Botanique* with the following statement: 'Botany, the science that deals with plants, has two parts that one must distinguish carefully: knowledge of plants, and knowledge of their powers (virtues)' (1694: 1, our translation from the French). For the French botanist, the inventor of that first part was the philosopher Theophrastus, while that of the second was the physician Hippocrates. Tournefort did not deny the utility of applied botany, but asserted that the more theoretical part must 'of necessity precede the study of plant virtues'. The Swedish botanist Carl Linnaeus (1707–1778), for his part, distinguished between botanists and plant collectors (1755: 4). In the early twentieth century, the American botanist and historian of the discipline Edward Lee Greene (1843–1915) gave the following definition of botany:

> In the most extended use of the term, all information about the plant world or any part of it is botany. According to this view, all treatises upon agriculture, horticulture, floriculture, forestry, and pharmacy, in so far as they deal with plants and their products, are botanical. What many will consider a better use of the term is more restricted. In this use of it there will be excluded from the **category of the properly botanical** whatever has no bearing on the philosophy of plant life and form.
>
> (Greene 1909: 7, our emphasis)

Greene then called these two categories 'botany' and 'plant industry' respectively, recognising that some traces of 'genuine botany' can occur in more practical texts. Agnes Arber (1879–1960), the Cambridge botanist, in her history of herbals talked about 'philosophical and utilitarian' botany, pointing out, however, the somewhat arbitrary nature of this division (1912: 1). The historian of science Charles Singer (1876–1960), in an important article on ancient herbals, wrote rather scathingly on the two branches of plant lore:

> A *Herbal* is a collection of descriptions of plants put together for medicinal purposes. Most herbal remedies are quite devoid of any rational basis. It may be taken for granted that the writer of a herbal is unable to treat evidence on a scientific basis. He makes a 'direct attack' on disease, without any 'nonsense about theories.' **The herbal is thus to be distinguished from the scientific botanical treatise** by the fact that its aims are exclusively 'practical' – a vague and foolish word with which, from the days of Plato to our own, men have sought to conceal from themselves and from others their destitution of anything in the nature of general ideas.
>
> (Singer 1927: 1, our emphasis)

No wonder the research into the history of pharmacology lagged behind that of other medical disciplines for so long! To give a final example – the list could go on – the classicist Reinhold Strömberg opens his detailed study of Theophrastus' botanical writings thus: 'The beginnings of the Greek research in nature can be divided into two branches: one pertains to natural philosophy and seeks through speculation to explain natural phaenomena; the other pertains to observation, and practically and empirically interprets natural objects' (1937: 17, our translation from the German).[2]

The problem with such a division is two-fold. First, it is not an 'actor's category'. The ancients would not have divided plant science in such a way; the utility of plants and the submission of the vegetable kingdom to mankind are at the basis of most ancient 'botanical' texts. Second, beside Theophrastus' writings (*Enquiry into Plants* and *Causes of Plant Phaenomena*) and Nicolaus of Damascus' *On Plants*, no ancient text focuses on what we call 'pure' or 'theoretical' botany. Theophrastus himself wrote on the topic of what we could term 'applied botany': his treatises *On Odours* (preserved) and *On Wine and Oil* (lost). Book nine of the *Enquiry into Plants*, devoted to pharmacological plants, is also quite practical and, as a result of the prejudice against 'practical botany' has often been considered spurious (see p. 9). If we focus too much on this theoretical side of ancient plant science, we run the risk of reaching the same conclusions as the historian of botany Robert Harvey-Gibson: 'After Galen follows an absolute blank; for more than fourteen centuries botany had no history. Theophrastus had to be rediscovered, or rather all that he taught had to be relearnt' (1919: 10). One of the aims of this book is to redress this claim. There is good botanical expertise in herbals and/or in late-antique texts. In sum, by 'botany' in this book we mean 'the technical knowledge of plants, their names and morphology (form), their classification, their physiology, and their habitats'. We argue that this technical knowledge was quite widespread in the ancient world.

The second reason for choosing this passage of Dioscorides is that he is dismissive of *historia*, that is, enquiry as contrasted to *autopsia*, personal observation. In other words, Dioscorides is criticising his predecessors for reading instead of doing fieldwork. Unlike these authors, Dioscorides claimed, he had acquired his botanical knowledge through experience and observation. Now, botanical experience and observation were not the sole preserve of specialist scholars; vital information on plants could be contributed by people at all levels of ancient society, be they woodworkers, country women or mighty kings. Many of these people were illiterate, but their observations and theories were written down by literate authors. At times, these authors may have misinterpreted, unwittingly or consciously, their oral sources. It is always difficult for today's historians to analyse ancient oral sources through the filter of written texts, but we will nevertheless attempt to point out traces of the ancient oral botanical tradition whenever possible. Nor will we limit our discussion to texts that deal solely with plants; we shall consider works by historians, poets and Jewish authors.

Our main aim, therefore, is to place ancient botany in the social, economic and cultural context of the Greek and Roman world. We choose to present our material thematically rather than chronologically, because there are already many chronological overviews available, although some of them are rather outdated. Among such overviews, we can mention those by the German botanist and physician Kurt Polycarp Joachim Sprengel, *Geschichte der Botanik* (1807–1808, in two volumes); that by the German botanist Ernst Heinrich Friedrich Meyer, also entitled *Geschichte der Botanik* (1854–1857, in four volumes, the two first volumes are devoted to the ancient and medieval world); that by the American botanist Edward Lee Greene, *Landmarks of Botanical History* (1909); and that by Liverpool-based botanist Robert Harvey-Gibson, *Outlines of the History of Botany* (1919). As is quite clear, the genre of the monumental history of botany has been dominated by botanists rather than historians.[3] Here we combine our classical/historical and botanical expertise to provide a thematic approach to ancient botany, although we do offer a – roughly – chronological overview of our main, *preserved*, sources in this Introduction. We will introduce other written sources that are not fully preserved, as well as sources that mention plants but do not focus on them, in Chapter 2 and throughout the remainder of this book. From a chronological point of view, we will go from the eighth century BCE (date of the writing down of the Homeric poems) to the seventh century CE (which saw the advent of Islam and new influx of botanical knowledge). We have taken into account as much of the secondary literature dealing with ancient plants as possible. It is very extensive and comes from numerous disciplines: history of science, medicine and technology (the branch of scholarship with which Laurence Totelin associates the most); botany (the branch of scholarship with which Gavin Hardy associates the most); classical studies, which deal with editing ancient texts on plants; history of philosophy, which deals with authors such as Aristotle and Theophrastus; ancient history, which studies the economic and social impact of plant exploitation, and the use of plants as foods; literary studies, which examine texts (in particular poems) that describe plants; archaeology, and in particular garden archaeology, which deals with remains of plants found in archaeological context; history of art, which deals with ancient depictions of plants; numismatics, which deals with coins, many of which bear plant depictions; codicology, which deals with manuscripts, some of which are botanical in nature; quaternary science, which deals with the environment of the quaternary period; ethnobotany and ethnopharmacology, which sometimes use history to understand current practices, or to develop new drugs. The study of ancient botany has recently made great advances, in particular with the edition, French translation, and extensive commentary of Theophrastus' *Enquiry into Plants* by French scholar Suzanne Amigues (1988a–2006); as well as the translation into English of Dioscorides' *Materia Medica* by Lily Beck (2005). Both these works are based on extensive analysis attempting to identify plants listed in ancient texts; they offer the most reliable identifications, although there is still much scope for discussion. Plant identification is often seen as an obstacle in the study of ancient botany. To a certain extent, this is the case, but we believe there is now enough data in this field to proceed to write a more general work on ancient botany, a work that does not have identification as its central concern. We therefore do not offer new identifications of plants here and refer the reader to the large body of literature on the topic (see Chapter 4 for more detail). We shall mention works identifying the plants mentioned by particular ancient authors where appropriate.

In the remainder of this Introduction, we present the main sources available for the study of ancient botany: textual sources mainly, but also material ones. In Chapter 2, we examine how the authors presented in Chapter 1 obtained their botanical knowledge. We identify three main types of sources: personal observation, oral sources, and written texts. As we discuss these sources, we will encounter various 'actors' in the field of ancient botany, people who handled plants on a daily basis, and who range socially from the lowliest of slaves to the highest of kings. Our section on written texts will allow us to introduce some sources which were not presented in Chapter 1, either because they have not been preserved, or because they were not focused solely on plants.

In Chapter 3, we tackle the issue of plant classification in the Graeco-Roman world. We start by asking how the ancients defined a 'plant', introducing ancient notions of the soul, and in particular Aristotelian notions. We then introduce the ancient notions of genus and species, as well as the various systems of organisation of plants in existence in antiquity. We pay particular attention to Theophrastus' organisation into trees, shrubs, undershrubs and herbs, traces of which are present in many other Greek and Latin texts.

In Chapter 4, we examine how plants were named, described and depicted in the ancient world. It is here that we discuss the thorny issue of identifications of plants named in Greek and Latin texts. Identifying those plants is by no means easy; however, we argue that ancient plant descriptions and depictions are much more valuable for that purpose than modern scholars often acknowledge.

In Chapter 5, we follow the life cycle of a plant from generation (which could take various forms), to growth, maturity and death. We also discuss annual phaenomena such as flowering, fruiting and leaf-shedding. We observe the prevalence of anthropomorphism in ancient descriptions of plant physiology. In particular, we note how plants were said to be male and female; to have sexual intercourse; to raise children; and to suffer the same illnesses as humans.

Our final chapter, Chapter 6, deals with the relation between plants and their environment. We note how ancient authors considered cultivated trees to be completely different from wild trees. We examine ancient stories of plant transplantation and briefly discuss marine plants. We end this chapter with some considerations about human impact on flora in antiquity.

1.2 The sources

1.2.1 Literary sources

Plants appear in Greek texts from the very beginning. Indeed, texts in Mycenaean Greek found at Pylos, Mycenae and Knossos (see Map 1), bear names of plants used as spices and in perfumes (see Foster 1974; Shelmerdine 1985).[4] The Homeric poems, the earliest texts copied in alphabetical Greek, also contain many references to plants (see Chapter 2 for more detail). We start our survey, however, in the fifth century BCE, with the medical writings attributed to Hippocrates.

a The Hippocratic Corpus

As we saw, Joseph Pitton de Tournefort identified Hippocrates as the creator of the practical branch of botany. Hippocrates was a physician from the island of Cos (see Map 1) active in the fifth century BCE. A body (corpus) of some sixty texts on various medical topics was attributed to him in antiquity. Scholars now agree that it is impossible to determine which ones (if any) of these treatises Hippocrates himself wrote. It has therefore become conventional to refer to the 'Hippocratic authors/writers/physicians' rather than to Hippocrates.[5]

The Hippocratic texts are among the earliest prose texts preserved in Greek, as they date for the most part to the fifth and fourth centuries BCE. Some of these texts are contemporary with the writings of the pre-Socratic philosophers, which are only preserved in fragments. Hippocratic authors shared many concerns with the pre-Socratic philosophers, and some Hippocratic texts can be classified as pre-Socratic philosophical texts. For instance, the Hippocratic author of *On Generation* and *Nature of the Child*, active at the end of the fifth century BCE (see Jouanna 1999: 392), was interested in embryology as were the pre-Socratic philosophers Empedocles (fifth century BCE) and Democritus (fifth–fourth century BCE). This Hippocratic author had a particular taste for plant analogies when discussing the development of the human child and included a long botanical excursus in his work (chapters 22–26, Loeb 10.60–78 Potter = 7.514–528; see Lonie 1969, 1981).[6] That excursus is our longest text on plant physiology written before Aristotle. Analogies between plants and humans are to be found in other texts of the collection.[7]

As physicians, the Hippocratic authors also made use of plants in pharmacological and dietetic preparations. The Corpus includes over 1500 recipes, most of which are based on herbal ingredients, and many dietetics prescriptions besides (see Totelin 2009). In total, approximately 300 plants are listed in the Corpus.[8] On the other hand, it does not contain any text on 'simple remedies', that is, a catalogue of pharmacologically active substances including botanical descriptions and medical indications (see pp. 21–22 for the treatise *Simples* by Galen). The Corpus, however, contains an important catalogue of foods, preserved in the treatise *Regimen* II (see

Wilkins 2004). The catalogue contains relatively little detailed information on the plants it lists (beside their dietetic properties), but it served as a model for later catalogues that were to include such information. We will discuss some of these catalogues later, but here is the place to mention the *Dynamidia of Hippocrates*, a late-antique Latin text based in part on a Latin translation of *Regimen* II, to which were added various pieces of information extracted from other sources – the *Dynamidia* have much in common with Isidore of Seville's *Etymologies* and with *Medicines made from Female Herbs*, falsely attributed to Dioscorides (see pp. 21 and 24–25 for both these texts). The Latin word '*dynamidia*' derives from the Greek word '*dunamis*', which refers to the medicinal powers of plants. The *Dynamidia of Hippocrates* appear to have included four books, but this text was very much 'alive', which means that copyist did not hesitate to add and remove information. As a result, no two manuscripts of the *Dynamidia* contain the exact same text, and no satisfactory modern edition is available. In this book, we have used the passages edited by Arsenio Ferraces Rodríguez in his study of various late-antique Latin therapeutic texts (1999).

Also preserved within the Hippocratic collection are letters allegedly sent by Hippocrates to various correspondents, including the philosopher Democritus. These letters are not authentic, that is, they were written well after the death of Hippocrates – they are what classicists call 'pseudepigraphic letters' (see Smith 1990). Letter sixteen is a letter allegedly sent by Hippocrates to Crateuas, the *pharmakopōlēs* (drug-seller), and includes some useful information on the collection of medicinal herbs (70–73 Smith = 9.342–348 Littré).

b Aristotle and pseudo-Aristotle

According to the philosophical biographer Diogenes Laertius (third century CE), Aristotle wrote a treatise *On Plants* in two books (Diogenes Laertius, *Lives of the Philosophers* 5.25).[9] While a work in two books *On Plants* is preserved in the Aristotelian Corpus, it is now accepted as the work of the Aristotelian philosopher Nicolaus of Damascus (first century CE, see pp. 11–12).[10] Aristotle refers on various occasions to a treatise on the subject of plants, sometimes in the past tense, which seems to indicate that he actually wrote this work and that it is now lost.[11] Scholars have long debated about this possibly lost treatise. Gustav Senn (1929) argued that, when he mentioned *On Plants*, Aristotle was in fact referring to the research of his student Theophrastus. Otto Regenbogen (1937) disagreed with this conclusion, showing that Theophrastus sometimes criticised his master in the field of botany. We will see that there are indeed differences between the thinking of Aristotle and Theophrastus, in particular in the use of analogies between plants and animals. When Aristotle alludes to his treatise *On Plants*, he is either referring to a lost treatise or to a treatise he planned to write, but never managed to complete.

If Aristotle wrote a treatise *On Plants*, it was lost by the second century CE. The Aristotelian commentator Alexander of Aphrodisias indeed wrote that 'there is a work *On Plants* written by Theophrastus. That by Aristotle is not preserved' (*Commentary on Sense and Sensible Objects* 442b25–26). Other Aristotelian commentators refer to Aristotle's *On Plants*, but it is unclear whether they actually saw this work, and what they say about it tells us almost nothing about its possible content.[12]

While it is not clear whether Aristotle ever wrote *On Plants*, it is certain that he had an interest in them, as he mentioned them in all his biological treatises (see Wöhrle 1995). The relevant passages have been conveniently collected by Friedrich Wimmer (1838). Aristotle was particularly interested in the position of plants in the great chain of being (*scala naturae*, see Chapter 3) and in their physiology. He left the description of actual plants species to his follower, Theophrastus.

Aristotelian material on the topic of plants is also to be found in the *Problems*, a series of thirty-seven books of questions and answers on topics ranging from human diseases to mathematical problems. Some of these problems may have originated in the Lyceum, but it is clear that Aristotle himself did not compile them.[13] Plants feature in many books of the *Problems*, and three books focus on them: books twenty (on shrubs and vegetables), twenty-one (on cereals, flour and bread) and twenty-two (on fruits). Book twenty bears much similarity to book five of Theophrastus' *Causes of Plant Phaenomena*, and this has led the French editor Pierre Louis to argue that *Problems* twenty is a genuine work of Aristotle, written around 340 BCE (Louis 1993: 120). This may be stretching the evidence a bit too far, but it is plausible that the ideas expressed in *Problems* twenty originated in Aristotelian circles, and might date to the fourth century BCE.

c Theophrastus of Eresus

Theophrastus, (*c.*372/371–287/286 BCE, see Sollenberger 2008 for references) originally from Eresus on the island of Lesbos (see Map 1), moved to Athens (see Map 1) as a young man to study under Plato and then Aristotle. He accompanied Aristotle on several of his travels, and may have been with him at the court of Philip II of Macedon. He became the head of the Lyceum while Aristotle was still alive (he had fled Athens where anti-Macedonian sentiments were rife), and remained its leader until his death (see Gottschalk 1998 on Theophrastus and the Peripatos).

According to Diogenes Laertius, Theophrastus 'acquired his own garden after the death of Aristotle. Demetrius of Phaleron, to whom he was acquainted, assisted him in this enterprise' (*Lives of the Philosophers* 5.39, see Sollenberger 1985: 14). The 'testament of Theophrastus' also preserved by Diogenes Laertius, mentions a garden alongside a promenade and a house to be used together, as if they were sacred ground. Theophrastus expressed the wish to be buried in a suitable part of the garden, and makes arrangements for the care of the garden after his death (*Lives of the Philosophers* 5.52–54; see Sollenberger 1985: 40–41). If this testament is genuine, it indicates that Theophrastus had a practical interest in the cultivation of plants.

Theophrastus was a prolific author – Diogenes attributes 225 works to him. Most of this oeuvre is philosophical in nature, but Theophrastus is also the author of the famous *Characters*, where he depicts various types of people, such as the flatterer or the reckless man. Much of Theophrastus' philosophy is lost, but most of his botanical works are fortunately preserved. These are *Enquiry into Plants*, in nine books, and *Causes of Plant Phaenomena*, in six books, to which we can also add *On Odours*. Numerous botanical fragments are also preserved in later works. Some of the fragments relate to works that are preserved today; others to lost works; and some are incorrect citations of Theophrastus' ideas.[14] In the botanical writings Theophrastus recorded numerous new observations and also summarised the views of many of his predecessors, including several pre-Socratic philosophers, whose works are not preserved.

Theophrastus' two major botanical works are quite different from one another and were meant to complement each other. According to the editor of Theophrastus, Suzanne Amigues, neither were meant for public circulation, but were rather notes accompanying oral lectures at the Lyceum. This would explain the elliptic style of both works (Amigues 2012: VII).[15]

The *Enquiry into Plants* is an observational work, which describes plants and habitats. It is similar to Aristotle's *Enquiry into Animals* in its aim of outlining the 'differences' (Greek: *diaphorai*; Latin: *differentiae*) of plants in a systematic manner (see Gotthelf 1988: 116). Book one and the first chapters of book two (2.1–4) deal with plants generally (*kath'holon*); Theophrastus then deals with plants individually (*kath'hekasta*), dividing them into four categories: trees, shrubs, undershrubs and herbs (see Chapter 3 of this book). The *Enquiry* ends with a book (book nine) on medicinal plants, whose authenticity has often been contested, some scholars arguing that it had been composed by an unknown root-cutter or that it was compiled from the writings of the physician Diocles of Carystus (fourth century BCE, see Manetti 2008 for references).[16] The current scholarly opinion, however, is that this book is authentic. Suzanne Amigues showed that book nine is in fact composed of two distinct treatises, both authentically Theophrastean. Chapters one to seven are a tract *On the Saps of Plants* (Greek: *Peri phutōn opōn*), dealing with aromatic plants. Chapters eight to nineteen, for their part, are a treatise *On the Power of Roots* (Greek: *Peri dunameos rhizōn*), dealing with pharmacologically active plants, also known as 'roots'. According to Amigues, neither *On the Saps of Plants* not *On the Power of Roots* were destined to be joined together, nor were there meant to become book nine of the *Enquiry*. In fact, both short treatises were written before the *Enquiry* (Amigues 1998, 2006: introduction).

Causes of Plant Phaenomena focuses on the 'causes' (*dia oti*) of plants, that is, the reasons why plants generate, grow, produce taste and odour, etc. It is both a textbook of plant physiology and a detailed handbook of agricultural methods. Its structure is rather complex. It starts with plant processes that are natural and spontaneous: their generation and growth (books one and two). It then turns to the interactions between man and plants (books three to five). Book six deals with the taste of plants. *Causes of Plant Phaenomena* may also have originally included a book on plant smells (a theme further developed in the independent treatise *On Odours*), and a book on wine and oil.[17] Theophrastus composed the *Causes* after the *Enquiry* and intended the two treatises to be used in conjunction. Indeed, he opens the *Causes* with the following statement: 'The modes of plant generation are several in numbers, as we have said earlier in the *Enquiries*, where we also enumerated and named them' (*CP* 1.1.1).

On Odours deals mainly with plant smells, but also includes information about animal smells. Georg Wöhrle comments that this treatise is not preserved in full – as is made obvious by its abrupt ending – but suggests that not much of the original text is missing since the most important aspects mentioned in the introduction are discussed in the text as it exists. He also argues that certain questions, such as the perfuming of wines, may have also been discussed in one of the putative missing books of *Causes*, the book on wine and oil (1988: 11).

d Nicander of Colophon and other learned Greek poets; Vergil

Following Theophrastus, the next major figure to appear in Greek botany is Nicander of Colophon. For many years scholars were uncertain where to place Nicander chronologically, but current thinking is that he flourished in the period 197–130 BCE at the time of Attalus III, the last King of Pergamum (see Thibodeau 2008a for references). As a poet, he appears to have worked in the entourage of Attalus III, who had an interest in plants, poisons and antidotes (see Scarborough 2010; Totelin 2012a).

Two works of Nicander are preserved: the poems *Theriaka* and *Alexipharmaka*, both in hexameters.[18] The former deals with venomous snakes, spiders and scorpions and the treatment of their bites; the latter with twenty-two poisonous plants, animals and minerals. The principal botanical input of Nicander comes from his *Alexipharmaka* which includes plant poisons such as aconite, hemlock and opium. The cures are almost always herbal and often include olive oil as an emetic to induce vomiting. Nicander also wrote treatises – now lost save for fragments – on a variety of other topics such as farming, geography, hunting and poets. Nicander's *Georgics* were also admired by Cicero (see *On the Orator* 1.69) and extensively quoted by Athenaeus (second century CE).

Scholars have differing opinions concerning both Nicander's poetic and pharmacological abilities. Nicander's verses are certainly full of witticisms and are often convoluted – which makes the task of identifying the plants and animals he depicts arduous – but that was a common characteristic of ancient technical poetry. As for his pharmacological skill, the dominant opinion used to be that Nicander had merely versified the iologic prose treatises of the physician Apollodorus (third century BCE, see Jacques 2008a for references), often showing his ignorance in the process (see in particular Schneider 1856: 181–201; Wellmann

1898a: 23–28; Gow and Schofield 1953: 23–25). Jean-Marie Jacques reappraised the poet's medicinal knowledge, and concluded that Nicander was both a good physician and poet (1979, 2008d; see also Touwaide 1991 for a nuanced analysis). John Scarborough (2012b) is more circumspect, stating that the court poet was 'neither zoologist nor toxicologist'. It remains that Nicander was reputed in antiquity for his pharmacological skills. For instance, Pliny mentions him as a source in the indices to books twenty to twenty-seven of the *Natural History*, the books that deal with pharmacology. The Byzantine encyclopaedia Souda talks of him as a grammarian, poet and physician (s.v. *Nikandros*, N374 Adler). Because Nicander wrote in verse, his statements were easy to memorise and, as a result, his *Theriaka* and *Alexipharmaka* were used by students of toxicology until the Renaissance, and a large number of scholia (explanatory notes) were produced to explain all their textual obscurities.

Nicander was not alone in writing extremely learned poems on plants. Many are lost, as for instance the *Garden of Health* by Nestor of Laranda (fl. *c.*195–210 CE, see Keyser 2008b for references), quoted by Cassianus Bassus, himself also only known second-hand through the Byzantine *Geoponika* (12.16.1 and 12.17.16–17, see pp. 14–15). Other such poems are preserved in fragmentary form, as is the anonymous poem *On Egyptian Plants* (*Anonymus de plantis Aegyptiis*), preserved on papyrus dating to the second century CE (P.Oxy. 15.1796 = Trismegistos 63593).[19] The twenty-two hexameters that are preserved concern cyclamen and persea. The entire poem appears to have dealt with the Nilotic countryside. Fragments (amounting to 216 hexameters) of a third-century *Song on the Virtues of Herbs* (*Carmen de viribus herbarum*) are also preserved in a handful of manuscripts (see p. 16 for the Vienna Dioscorides).[20] It deals with seventeen medicinal plants and is quite magical in tone.

The most famous ancient such poem, however, is Vergil's *Georgics* in four books of Latin hexameters. P. Vergilius Maro (70–19 BCE, see Thibodeau 2008g for references) was an orator acquainted with philosophy, in particular Epicurean philosophy. Under the rule of Augustus, Vergil was patronised by Maecenas, patron extraordinaire of the arts. His poems include the *Bucolics*, the *Aeneid* and the *Georgics*, which interests us here. Book one is devoted to cereals and weather signs; book two to the vine and fruit trees; book three to animal husbandry; and book four to apiculture. Book four includes a short excursus on horticulture (4.116–148), which would be the starting point for the poems of Columella and Palladius (see pp. 13–14).[21] In total the *Georgics* name 164 plant species and demonstrate a certain first-hand knowledge of botany, although for the most Vergil derived his information from the writings of his – mostly Greek – predecessors.[22]

e Nicolaus of Damascus

Nicolaus of Damascus was a polymath from Judea (see Map 3). He was counsellor to King Herod the Great before settling in Rome (see Map 4) at the court of

Augustus (see Zucker 2008b for references). Much of his works are lost, and he is now mostly remembered as a commentator of Aristotle. In the field of plant science, he composed *On Plants* (*De plantis*) in two books. This was for a long time thought to have been written by Aristotle. In fact, *On Plants* is a compilation of Aristotelian and Theophrastean material with additions by Nicolaus himself.[23] Book one deals with the question of the soul in plants and the parts of plants; it then covers material very similar to book one of Theophrastus' *Causes of Plant Phaenomena*. In book two, Nicolaus deals with plant physiology, and the places where plants grow, with a long excursus on salt water.

The transmission history of Nicolaus' *On Plants* is extremely complex: it was translated and abridged into Syriac in the ninth century, but only fragments of that translation exist (on this transmission, see Drossaart Lulofs 1957; Drossaart Lulofs and Poortman 1989: 1–14). The Syriac version was then translated into Arabic around 900 by Ishaq ibn Hunayn, as the extremely informative title of the translation indicates: 'the treatise on plants by Aristotle: an adaptation of Nicolaus. Translated by Ishaq ibn Hunayn with corrections by Thabit ibn Qurra' (Drossaart Lulofs and Poortman 1989: 126). Since the Syriac translation is so little known, the Arabic version is the central text – it is the one we will use here. The Arabic translation was then rendered into Hebrew and into Latin by Toledo-based translator Alfred of Sareshel in the thirteenth century CE (see Meyer 1841; see Map 3). The Latin translation circulated widely in the Middle Ages (around 150 manuscripts) and was commented upon by Albert the Great (thirteenth century), among others. In the Renaissance that Latin translation was finally rendered into Greek.

The text as we have it is rather confused, which is partly due to its complex transmission and partly to Nicolaus' method of compilation: he drew both on Aristotle and Theophrastus, adding his own views, albeit in a somewhat rambling summarised format with an abrupt ending. In the words of his editors:

> Nicolaus had done his best to convey the outline of a long, intricate and highly complicated text in the smallest possible space. His efforts were bound to fail, and they resulted in a rather incoherent series of diverse information on botanical subjects.
>
> (Drossaart Lulofs and Poortman 1989: 13)

Nevertheless, some sections are clearly lifted from Theophrastus without any ambiguity, for example the section of text which classifies plants on the basis of whether they are trees, shrubs, undershrubs or herbs (see Drossaart Lulofs and Poortman 1989: 104).

f The Roman agronomists: Cato, Varro, Columella, Palladius

Our next sources are the Roman agronomical writers: Cato the Elder, Varro, Columella and Palladius.[24] They were not primarily interested in plant science, but they show themselves extremely knowledgeable, in particular – and not surprisingly – in the fields of agriculture, horticulture, pomiculture and viticulture.

Marcus Porcius Cato (234–149 BCE, see Thibodeau 2008d for references) was a member of an old plebeian family. He spent his childhood on his father's farm near Reate (Sabine region; see Map 4). He had distinguished military and political careers, which brought him to various places: Sicily, Africa, Sardinia, Spain and Greece. He is best known for his criticism of Greek culture and the Hellenisation of Roman ways. He was also a prolific writer, active in the fields of history, philosophy and law. His only surviving work is *On Agriculture* (*De re rustica*), the first prose work preserved in Latin.[25] It is a collection of agricultural advice and precepts, whose organisation is sometimes difficult to perceive, but based on first-hand experience. It starts with a preface in which Cato extols the qualities of the brave farmer and then covers topics such as how to choose a farm, equip and man it, plant and build it. It includes numerous recipes and pieces of practical advice.

Marcus Terentius Varro (81–27 BCE, see Thibodeau 2008e for references) was also from the town of Reate. His teachers included the philologist L. Aelius Stilo at Rome and the philosopher Antiochus of Ascalon at Athens. Like Cato, he had a political career as a partisan of Pompey the Great; he served as pro-quaestor of Pompey in Spain, of which he shows knowledge in his *Res rustica*. After Pompey's defeat in the Civil Wars, he was granted pardon by Caesar and asked to establish a public library at Rome – a project that was never brought to fruition. After Caesar's death, Varro got caught in the fight between Marc Antony and Octavianus (the future Augustus). He was condemned to death, but escaped thanks to powerful friends, including Octavianus. He devoted the rest of his life to scholarship, composing some 74 works, totalling 620 books, on topics such as geometry, astronomy, music and geography. Of this major output, only six books *On the Latin Language* (*De lingua Latina*) and the three books of *On Agriculture* (*Res rustica*) have survived. Varro composed it towards the end of his life (when he was eighty), and dedicated it to his wife, Fundania, who has just bought an estate (*On Agriculture* 1.1.1). Each book purports to be a dialogue between Roman landowners, who show great knowledge of past writings. The first book deals with the farm, the soil, equipment, plant crops; the second with animal husbandry; and the third with the breeding of birds, bees and fish. Throughout, Varro shows great interest in etymology and numerology (he endlessly divides his topics into sub-categories). He talks of agriculture as both an art and a science (*RR* 1.3–4), whose aims are profit and pleasure. Varro shows great knowledge of Theophrastus' works, but indicate that they are not particularly well adapted to practical ends (*RR* 1.5.1).

Very little is known of the third Roman agronomist: Lucius Junius Moderatus Columella (first century CE, see Rodgers 2008c for references). He tells us he was from Gades (southern Spain, see Map 3), and spent much of his youth in the company of his uncle Marcus Columella, a careful farmer. At some point he left Spain to settle near Rome, where he purchased various farms. He also travelled to Syria and Cilicia (see Map 3). He composed his *On Agriculture* (*Res rustica*) in twelve books and *On Trees* (*De arboribus*) in one book.[26] *On Agriculture* is the most systematic of the Roman treatises on agriculture; it is dedicated to a certain Publius Silvinus – perhaps a fictional character. The treatise opens with the choice of land for a farm, its building and staff (book one); followed by ploughing and the care of the crops (book two); the care of fruit trees, vine and olive (books three to five); care of farm animals, starting with the biggest (cattle, horses and mules: book six), down to the smallest (bees: book nine). Book ten is different in form (see Marshall 1918 for an introduction to this book).[27] It is a poem in hexameters on gardening, a topic that was, apparently, novel at Rome: 'Then it remains to deal with the cultivation of gardens, a topic which the husbandmen of old lazily neglected, but which is now much celebrated' (Columella, *On Agriculture* 10.1.1). His model for book ten was Vergil's *Georgics*: '[You repeatedly asked me] to complete in poetic style those parts of the *Georgics* which Vergil omitted, and which as Vergil himself indicated, he left to his successors to treat' (*On Agriculture* 10.1.3). Book ten was intended to close the work, but at the insistence of Silvinus (11.1.1), Columella added on the duties of the overseer, a calendar and a long passage on gardening, storage of vegetables, herbs, cheese, wine and olive oil (books eleven and twelve), and a book on the duties of the overseer's wife (book thirteen). Throughout *On Agriculture*, Columella demonstrates great knowledge and expertise. The independent treatise *On Trees* deals with the vine, the olive and other fruit trees, and contains much that is also to be found in *On Agriculture*. Columella remained a respected agronomical authority throughout antiquity. He is cited as a source by Pliny and Palladius.

Palladius Rutilius Taurus Aemilianus (fourth or fifth century CE, see Rodgers 2008d for further references) is the fourth Roman agronomist. Again, we know very little about his life besides what he tells us in his work: *Work of Agriculture* (*Opus agriculturae*). He possessed farms in Sardinia (*OA* 4.10.16, see Map 3) and near Rome (*OA* 3.25.20). Like Columella, he often refers to his own expertise. His *Work of Agriculture* contains fourteen books: after a first general book, the twelve next books deal with the agricultural tasks to be carried out each month; book fourteen is devoted to veterinary medicine.[28] The simple organisation by calendar month explains the popularity of his work in the Middle Ages. To this work in prose, Palladius appended a poem in elegiac couplets (hexameters alternating with pentameters) *On Grafting* (*De insitione*) dedicated to the unknown Pasiphilus. Fruit trees occupy the largest portion of Palladius' works. On this topic, Palladius may have borrowed some information from Gargilius Martialis (third century CE), who wrote a treatise on gardens and orchards (*On Gardens* (*De hortis*)), only known through fragments (edited by Mazzini 1988).[29]

Here is also the place to mention the *Geoponika*, a twenty-book Byzantine Greek farming handbook of the tenth century CE, dedicated to the Emperor Constantine Porphyrogenitus.[30] Although it falls outside our chronological span, it is of interest to us because it preserves extracts from sources now lost, or only known through translations in Syriac and/or Arabic. Thus, the compiler of the *Geoponika* appears to have borrowed heavily from the *Georgic Compilations* of

Cassianus Bassus Scholasticus (sixth century CE, see Rodgers 2008a), which in turn borrow the *Collection of Agricultural Practices* of Vindonius Anatolius of Beirut (fourth century CE, see McCabe and Rodgers 2008).

g Dioscorides and pseudo-Dioscorides

Apart from what he tells us in his *Materia Medica* (*On the Preparation, Properties, and Testing of Drugs*, known by its Latin title, *De materia medica*), we know very little about Dioscorides of Anazarbus (in Cilicia, modern Turkey).[31] He grew up in a Greek-speaking world fully under the control of Rome. He dedicated his work to Arius of Tarsus, a famous first-century CE physician active in the field of pharmacology, who appears to have been his teacher (see Scarborough 2008c for references). He tells us that he travelled widely thanks to his 'soldier's life' (preface 4), an allusion that has caused much debate among scholars. Some have read this as an indication that Dioscorides served in the army as a long-standing medical physician (see e.g. Wellmann 1903: 1131). The historian of medicine John Riddle, on the other hand, refused to believe Dioscorides had ever served in the army, because most of the places listed in the *Materia Medica* are important trading localities, and are situated in senatorial provinces rather than in the imperial provinces in which the military was based. He therefore proposed to interpret 'soldier's life' as a reference to a life filled with travels.[32] The historians of medicine John Scarborough and Vivian Nutton (1982: 213–217), for their part, argued that Dioscorides had been active in the army, but only for a short time, probably for the Armenian War – a conclusion later accepted by Riddle (1985: 4). The mention of Laecanius Bassus, consul in 64 CE and proconsul to the province of Asia in 79–80 CE in the preface (paragraph 4), with other clues studied by Riddle, indicate that Dioscorides wrote the *Materia Medica* around 60–80 CE (1985: xvii).

Dioscorides was a pharmacologist and was interested in all pharmacological substances. This consisted mainly of plants, but also included drugs derived from minerals and animals. For example, a cure for asp bites is to eat bedbugs (*MM* 2.34, see Riddle 1985: 138). The five books of *Materia Medica* deal with the following subjects: Aromatics, oils, ointments and trees (book one); animals, milk and dairy produce, cereals and sharp herbs (book two); roots, juices and herbs (book three); herbs and roots (book four); and vines, wines, and metallic ores (book five). Within each book, the substances are organised by what Riddle called 'drug affinities', that is, their effect on the body (see Chapter 3 for more detail). A chapter is devoted to each natural substance. In the case of plants, the chapter contains all or some of the following information, as discussed by Riddle (1985: chapter 2): 1) name of plant, and synonyms; 2) habitat; 3) botanical description; 4) pharmacological properties; 5) pharmacological uses; 6) possible side effects; 7) dosages; 8) information on how to collect, store and prepare the drug; 9) ways in which the plant is adulterated; 10) uses in veterinary medicine; 11) other uses (including magical ones).

The *Materia Medica* was, to use the phrase of Scarborough, 'an instant best seller' (2011: 7). It exerted a profound influence on pharmacological writing throughout antiquity and beyond. The text of the *Materia Medica* has a very rich transmission history. Versions following Dioscorides' original organisation as well as alphabetised versions circulated. Thus, books eleven, twelve and thirteen of the *Medical Collection* of Oribasius, the physician to the Emperor Julian, are a condensed and alphabetised version of Dioscorides' *Materia Medica*, introduced with the following words 'Book eleven: extracted from the writings of Dioscorides on the properties of simple remedies, following the alphabetical order, from alpha to mu' (CMG 6.1.2, p. 80 Raeder; see Scarborough 1984: 221–224).

The most famous manuscript containing the alphabetised version is the superbly illustrated 'Vienna Dioscorides' (MS Vienna, Österreichische Nationalbibliothek medicus graecus 1; see Zotter 1980 for a facsimile, introduction, transciption and German translation). The alphabetised version has here been shortened, but passages from Galen and Crateuas have been added, as well as numerous synonyms (see Chapter 4; Riddle 1985: 181). The 'Vienna Dioscorides' manuscript was dedicated in 512 CE to Julia Anicia, the daughter of Anicius Olybrius, Emperor of the West, by the inhabitants of Honorata, a district of Constantinople (see Map 3), in thanks for building a church (see Kiilerich 2001; Brubaker 2002). Julia Anicia is represented on one of the title pages, seated between Magnanimity and Prudence (fol. 6v, see Figure 1.1). Two other front pages of the manuscript represent botanical processes, from the plant discovery to its description and depiction: on folio 4v, Dioscorides is shown with Discovery (Euresis) holding a mandrake over a sick dog (see Figure 1.2); and on folio 5v, Dioscorides is presented writing in a codex, while Intelligence (Epinoia) holds a mandrake for a painter to depict (see Figure 1.3).

The manuscript contains other texts on plants and animals:

Folios 12–387: Dioscorides' *Materia Medica*, organised alphabetically, and with various additions.
Folios 388–392: *Song on the Virtues of Herbs* (see p. 11).
Folios 393–437: Paraphrase of Nicander's *Theriaka* by a certain Euctenius.
Folios 438–459: Paraphrase of Nicander's *Alexipharmaka* by Euctenius.
Folios 460–472: Incomplete paraphrase of Oppian's second-century poem *On Fishing* (*Halieutika*).
Folios 474–485: Incomplete paraphrase of the otherwise unknown Dionysius' poem *On Birds* (*Ornithiaka*).[33]

This manuscript has a complex history and has been annotated by various hands in Latin, Arabic and Hebrew.

Figure 1.1 Julia Anicia between Maganimity and Prudence, MS Vienna, Österreichische Nationalbibliothek medicus graecus 1 (512 CE), fol. 6v. Courtesy of Österreichische Nationalbibliothek, Vienna.

Dioscorides' *Materia Medica* was translated into Latin at least three times in late antiquity. Following the studies of Arsenio Ferraces Rodríguez, scholars have called these translations A, B and C (1999; see also Barbaud 1994; Cronier 2010). The first is only known through fragments and may go as far back as the third century CE. Passages thereof may be read in *Medicines made from Female*

Figure 1.2 Dioscorides with Discovery, who is holding a mandrake over a sick dog, MS Vienna, Österreichische Nationalbibliothek medicus graecus 1 (512 CE), fol. 4v. Courtesy of Österreichische Nationalbibliothek, Vienna.

Herbs (see p. 20). Ferraces Rodríguez attributes translation B, which is very poorly known, to the medical writer Caelius Aurelianus (fifth century CE, see Scarborough 2008e for references). It appears to have been a source for pseudo-Apuleius' *Herbarius* and for Isidore of Seville's *Etymologies* (see p.20). The

Figure 1.3 Diocorides writing in a codex, while Intelligence holds a mandrake for a painter to depict, MS Vienna, Österreichische Nationalbibliothek medicus graecus 1 (512 CE), fol. 5v. Courtey of Österreichische Nationalbibliothek, Vienna.

third translation (C) is the first Latin translation preserved in full. It is sometimes called 'Dioscorides Longobardus' (in reference to a manuscript) and dates to the sixth century CE. That translation was then alphabetised in the eleventh century, and additions were made (see Cronier 2010).

There also circulated under the name of Dioscorides a Latin late-antique treatise entitled *Medicines made from Female Herbs* (*Ex herbis feminis*).[34] It contains seventy-one chapters drawing heavily on Dioscorides' *Materia Medica.* It dates to the fifth or sixth century CE, and often circulated in manuscripts that contain the pseudo-Apuleius' *Herbarius* (see pp. 24–25). The title *Medicines made from Female Herbs* is somewhat mysterious. It was common in antiquity to refer to 'male' and 'female' plants (see Chapter 5), but this text does not deal exclusively with 'female' plants. It is possible that the treatise was originally conceived as dealing with women's herbs rather than female herbs, but that copyists misunderstood the title of the works (see Riddle 1981: 47–51). As already mentioned, chapters of *Medicines made from Female Herbs* derive from an early translation of Dioscorides' *Materia Medica* into Latin, but other chapters bear more similarities to chapters of pseudo-Apuleius' *Herbarius* (see Ferraces Rodríguez 1999: chapter 8).

h The encyclopaedists: Pliny the Elder and Isidore of Seville

Pliny the Elder (Gaius Plinius Secundus, 23–79 CE) was a member of the wealthy equestrian class. He took part in various military campaigns, where he met his friend, the future Emperor Vespasian. Under the rule of Vespasian, he had a political career as a procurator, an imperial adviser, and as commander of the Roman fleet at Misenum (Bay of Naples, see Map 4). He died while trying to observe the aftermath of the Vesuvius' eruption from a boat, as his nephew, Pliny the Younger, narrates in one of his letters (6.16). Pliny was a prolific author whose *Natural History* (*Historia naturalis*), an encyclopaedia in thirty-seven books, is preserved. It is dedicated to Titus, the son of Emperor Vespasian and his co-ruler. Recent scholarship has stressed the links between Pliny encyclopaedic project and Roman imperialism (see e.g. Murphy 2004; Beagon 2013). The preface is followed by a highly innovative Table of Contents, in which Pliny details the contents of each of the thirty-seven books and gives his sources, which he divides into 'Roman' and 'foreign' (see Doody 2001; see also Riggsby 2007 for tables of contents more generally). Pliny did not, however, consult all the sources he lists – his knowledge of many sources was second-hand. This is not to belittle Pliny's accomplishment, which remains monumental.[35]

In books twelve to sixteen, Pliny deals with trees and shrubs that grow wild; in books seventeen to nineteen, he discusses plants that are grown on farms and farm management; in books twenty to twenty-three he lists the dietetic properties of plants; while in books twenty-four to twenty-seven he discusses their pharmacological properties (see Chapter 3 for more detail). Pliny has often been considered uncritical and lacking in originality, and it is true that at times in the field of botany, he merely copies his sources, adding mistakes of his own. However, in places he shows himself very knowledgeable on plants, and adds personal observations (see Chapter 2).[36]

One should also mention here the encyclopaedia of Isidore of Seville (late sixth/early seventh century CE), the *Etymologies* (*Etymologiae*) in twenty books (see Somfai 2008 for references). This encyclopaedia focuses, as its title indicates, on the etymology of words, as Isidore thought that names provide the key to explaining Nature. Book seventeen is most relevant for us, as it covers agronomy and plants.[37]

i Galen, pseudo-Galenic texts, Oribasius and Aetius

Galen (129–*c*.216 CE) was born in Pergamum (modern Turkey, see Map 3), a cultural and medical centre at the time. He travelled widely to study medicine and spent most of his adult life in Rome, where he served under several emperors: Marcus Aurelius, Commodus, Septimius Severus and Caracalla (for a biography of Galen, see Boudon-Millot 2012). He was one of the most prolific authors of antiquity. Among his some 400 medical treatises, many are devoted to pharmacology and dietetics. Here we will be most concerned with *Mixings and Powers of Simple Drugs* and *Properties of Foodstuffs*.[38]

Properties of Foodstuffs (*De alimentorum facultatibus*, we will at times refer to this text as '*Foodstuffs*') is a dietetic treatise in three books dealing with cereals and pulses (book one); vegetables (book two); and animals (book three).[39] Therein Galen refers to Theophrastus on botanical questions, but sometimes adds interesting pieces of personal observation. There were many antecedents to Galen's foodstuff treatise, not the least the Hippocratic *Regimen* II and Cato's observations on the dietetic properties of certain plants, in particular the cabbage, at the end of *On Agriculture* (see in particular chapters 156–157). The world, however, in which Galen was writing was very different from that of classical Greece or second-century BCE Italy: Galen lived in an immense empire, with strong commercial links with India, Arabia and places even further. Therefore the array of plants included in *Foodstuffs* is much broader than that found in any preceding dietetic treatise. In the fourth century CE, Oribasius offered a summary of Galen's *Foodstuffs* in books one and two of his *Medical Compilation*, adding some passages from the works of other medical authors and from Athenaeus (CMG 6.1.1, pp. 4–65 Raeder, see Grant 1997 for a translation of book one).

Mixings and Powers of Simple Drugs (*De simplicium medicamentorum temperamentis ac facultatibus*; we will refer to this text as '*Simples*') is a monumental work on simple medicines, that is, the powers of single ingredients, in eleven books. In the first five books, Galen expounds his theory of drug powers, and from book six, he describes individual ingredients. Books six to eight deal with plants (over 400 in total), organised in alphabetical order.[40] Galen, always a severe critic, acknowledged Dioscorides' work to be the best of its kind and manifested his respect by numerous citations (see Chapter 2 for more detail). However, Galen did not retain his predecessor's organisational principle by drug affinities, choosing instead the alphabetical order. Although his knowledge of plants is not as impressive as that of Dioscorides, Galen sometimes gives us interesting pieces of personal observation. He also has a particular interest in plant nomenclature. Indeed, Galen was also a keen linguist and wrote various lexica; in his *Lexicon of Hippocratic Terms* (*Linguarum seu dictionum exoletarum Hippocratis explicatio*), he explicated various plant terms.[41] In the sixth century CE, Aetius of Amida, possibly physician to the Emperor Justinian (see Scarborough 2008a for references), offered a summary, with some additions, of Galen's *Simples* in the first two books of his *Medical Collection.* Aetius's book one is that devoted to plants; it lists 418 in total (CMG 8.1, pp. 17–146 Olivieri, see Scarborough 1984: 224–226). In the seventh century CE, the medical author Paul of Aegina, also offered his own concise summary of Galen's *Simples*, also with some personal additions, and also in the alphabetical order (*Epitome* 7.3, CMG 9.2, pp. 186–274 Heiberg; see Scarborough 1984: 228–229).[42]

Beside these authentic works by Galen, several pseudepigraphic treatises are relevant for a study of ancient plants. 'Pseudepigraphic' texts are treatises that were attributed to Galen from early on, but which cannot have been written by the physician (because their style is completely different from that of Galen or because they contain blatant anachronisms). One could mention, for instance, the *Alphabet of Galen* (*Alfabetum Galieni*), a text (in Latin) covering some 300 pharmacological substances (including 220 plants), whose date of composition is almost impossible to determine. The *Alphabet* circulated widely in Europe until the thirteenth century (see Everett 2012 for an edition, translation and introduction to this text). The *Alphabet* is quite similar to Dioscorides' *Materia Medica*, but as argued by Nicholas Everett, the compiler of the *Alphabet* was not familiar with the text of Dioscorides (2012: 81). Rather the *Alphabet* and the *Materia Medica* must have made use of the same source(s), which may have included the lost works of Sextius Niger (first century CE, see Scarborough 2008l for references).

Another noteworthy pseudo-Galenic text is a short treatise *On the Virtue of Centaury* (*De virtute centaureae*), which only survives in a Latin translation from the Greek (by Niccolo da Reggio, 1341). The author of the tract is a physician who has come to Rome to practice, and has learnt about the many virtues of centaury from a doctor named Apollonius, and wants to teach them to his brother Pappias, another doctor. Vivian Nutton has recently studied this text (2010), and has pointed to the importance of the following sentence: 'My brother Pappias, I have seen in the case of the centaury what the famous teacher Themistius said about *arnoglossa*, the plantain, namely that…' Nutton argues (2010: 217–219) that 'Themistius' is a copyist's mistake, and that we should instead read 'Themison', the name of a famous first-century BCE physician (see Scarborough 2008m for references). Themison belonged to the Methodist sect, that is, a sect which focused on patient treatments rather than on complex medical theories. If we accept Nutton's hypothesis, *On the Virtue of the Centaury* would be one of the rare Methodist texts still extant.

j Astrobotanical texts and the Cyranides

Many ancient texts dealing with plants mention the importance of the seasons, rising of the stars, and other meteorological phaenomena in understanding vegetable growth. Astrobotanical treatises take this interest to the next level by assigning plants to heavenly bodies (sun, moon, and each of the five planets), and zodiac signs or zodiac decans. They work on the principles that links of sympathy or antipathy exist between various parts of the cosmos. Within the cosmos, affinities (sympathies) can exist between plants and parts of the body; plants with an affinity with the heart, for instance, will be particularly good remedies for that organ. Conversely, if there exists an antipathy between a plant and a part of the body, that plant will be poisonous for that part of the body. The concepts of sympathy and antipathy became very popular among Neoplatonic philosophers such as Proclus (412–485 CE).

The astrobotanical texts have recently been studied by the classicist Guy Ducourthial (2003), but there remains much work to be done on this rich material. The longest astrobotanical text is *On the Virtues of Herbs* (*De virtibus herbarum*, not to be confused with the poetic *Song on the Virtues of Herbs*) in two books, which survives both in the original Greek and in a Latin translation.[43] Book one deals with plants associated with zodiac signs (twelve plants), and book two with plants of the stars (seven plants). Medicinal recipes are given for each plant. In several manuscripts, this treatise is prefaced by a letter from Harpocration (an authority named in the *Cyranides*, see below) or Thessalus (the name of a famous Methodist physician, see Scarborough 2008n for references) to the Emperor Caesar Augustus, where he tells the story of his revelation in Egypt (see Chapter 2 for more detail).[44] In either case, the attribution is most certainly pseudepigraphic, and the date of the treatise is therefore almost impossible to determine, with scholars giving a wide range of possible dates from the first century CE to the fourth.[45]

Other, shorter astrobotanical treatises have survived. They are scattered in various publications and can be difficult of access.[46] They are almost impossible to date. Ducourthial concludes that they cannot antedate the Hellenistic period, and that most postdate the second century CE. They are attributed to various famous authors: the Macedonian King Alexander the Great, the wise Jewish King Solomon, Hermes (see below), the Egyptian high priest Petosiris, etc. These people, when they are historical characters (not all are), most certainly never wrote these texts, but their names conferred authority on this astrobotanical material. One such short astrobotanical text is *On the Properties of Peony* (*Virtutes herbae paeoniae*), which circulated both in Greek and in Latin, and which deals with the magical and therapeutic properties of the peony. It exists in several versions, some of which were written in a Christian context. It contains some short botanical indications on the environment in which peony grows, and the seeds of the plants (see Ducourthial 2003: 298–303).

Botanical material is also included in the Hermetic *Cyranides*, a Greek treatise whose transmission is very complex.[47] Book one is of most interest to us. It starts with a rather difficult preface, stating that the mythical Trismegistus received two works on antipathy and sympathy, one by Cyranus, King of Persia, and the other by Harpocration of Alexandria, who dedicated the work to his daughter. Hermes blended these two works and made a third, the current book one of the *Cyranides*. Book one as we have it contains twenty-four chapters, one for each letter of the Greek alphabet, each dealing with a plant, bird, stone and fish, and their medicinal/magical properties. Each chapter gives botanical, zoological and geological information, as well as numerous recipes. Book two deals with the medicinal/magical uses of land animals; book three with birds; book four with fish; book five, which is not fully preserved, deals with plants, and has much in common with the writings of Dioscorides; book six, of which only a few lines survive, deals with stones. Some of the material in the *Cyranides* may go back to the first centuries of the Common Era, but the compilation as we have it must date to the Byzantine period.

The treatise of pseudo-Plutarch *On Rivers* (*De fluviis*) is also worth mentioning for displaying a theory of universal sympathy. It is attributed to Plutarch, but was most certainly not written by the Greek-writing historian and philosopher (first–second century CE). Each chapter (twenty-five in total) presents a river, to which are associated stones and plants, often with magical and medical properties.[48]

k The Herbarius *anthology and the herbal of pseudo-Apuleius*

Numerous medieval manuscripts (fifty at least, sixty with fragments), dating mostly from the ninth to the thirteenth century, contain a collection of Latin pharmacological texts which we will refer to as the '*Herbarius* anthology'.[49] It gathers together various pharmacological texts (referred to as herbals in the literature) on plants, animals and minerals. These texts have various dates, from the third to the sixth century CE. Its longest text is the *Herbarius* (herbal) attributed to Apuleius. Apuleius was a philosopher and rhetorician active in the second century CE (see Voigts 1978). It is unlikely that Apuleius wrote the *Herbarius* – the compiler of the text used the name 'Apuleius' as he thought it would confer authority to his writings, make them more worthy of attention. The editors of the text suggest that is was written in North Africa in the fourth century CE (Howald and Sigerist 1927: xx). Throughout the centuries, additions were made to the text: new plants, new recipes, and in particular the addition of plant synonyms. The *Herbarius* contains 131 chapters, each devoted to a separate plant and including recipes to treat various ailments, organised in the 'head-to-toe' order. Botanical information, including numerous plant synonyms and information on habitats, is found at the end of the chapters.

The other texts included in the *Herbarius* anthology are as follows:

1 Prayers (*Precationes*) to be recited while gathering herbs or preparing medications. The two prayers here included are the *Prayer to the Earth* (*Precatio terrae*), addressed to the goddess Earth; and the *Prayer concerning all*

Herbs (*Precatio omnium herbarum*).[50] Numerous other prayers of that sort were composed in antiquity, some of them in a Christian context.

2 *Letter of Hippocrates to Maecenas*: a letter allegedly written by the famous physician and including medical information on ailments and dietetic treatments. It was probably originally written in Greek then translated into Latin.

3 Antonius Musa, *On Betony* (*De herba vetonnica*): a short text on the numerous medical uses of the herb betony. It contains numerous uses for this important plant. Again, the name 'Antonius Musa', that of the famous physician of the Emperor Augustus (see Scarborough 2008b for references), was used to confer authority to this text. This text belongs to the same tradition of short treatises devoted to a single plant as *On the Virtue of Centaury* and *On the Properties of Peony*. Vivian Nutton notes that: 'Such a type of literature, detailing over several pages the medicinal uses of a single plant, is likely to have been an early casualty of the process of copying and recopying of manuscripts, for its practical applications could doubtless be found more briefly elsewhere in other more useful treatises' (2010: 215).[51]

4 Pseudo-Dioscorides, *Medicines made from Female Herbs* (see p. 20).

5 An anonymous text *On the Badger* (*De taxone liber*): short text on the medicinal uses of the badger.

6 Sextus Placitus, *Drugs from Animals and Birds* (*Liber medicinae ex animalibus et avibus*): a treatise on the medicinal uses of animals.

Some copies of the *Herbarius* anthology are illustrated, others are not. Illustrations include author portraits, medical scenes, and plant and animal representations. These illustrations have been studied in particular by Grape-Albers (1977) and Collins (2000). Peter Murray Jones (1999) introduces a facsimile of the thirteenth-century copy of the anthology in MS, Vienna Österreichische Nationalbibliothek 93, which contains stunning illuminations.

1.2.2 Plant representations

The illustrations included in the *Herbarius* anthology lead us to our next type of source: plant representations. It is not the aim of this book to study all representations of plants in Greek and Roman material culture. Suffice to say that they are omnipresent: ivy, vine, palm tree, myrtle on classical Greek vases (see Eichberger *et al.* 2007); laurel on cameo portraits of Livia, the wife of Augustus (see Flory 1955); fruits, vegetables and wheat on Hellenistic and Roman sculptures of the goddess Tyche; and so on and so forth. In Chapter 4, we shall study a selection of such plant representations, which vary in quality and in aim. Some aim to represent plants in a naturalistic way, while others are schematic. Those that are schematic may be so for various reasons: in the context of a technical text on plants, a schematic representation may be thought to be more appropriate than a naturalistic one that will, of necessity, only represent a plant at a particular stage of its development. On a coin, schematism is a necessity due to the small format of the medium. Many ancient plant representations are symbolic. There was in antiquity, as in later periods, a rich 'language of flowers'. In particular, plants were representative of human and animal sexuality and fertility.

1.2.3 Archaeological remains

Many plant representations were found in archaeological context (coins; sculptures; papyri). Archaeology is therefore an important source of information on ancient plants. We are interested here most particularly in one branch of archaeology, garden archaeology (see Miller and Gleason 1994 for an introduction to the discipline), pioneered by American archaeologist Wilhelmina Jashemski at Pompeii and the other Vesuvian sites (see Map 4).[52] Jashemski argued that:

> Archaeological evidence is extremely important, for it is not always possible to know exactly what plant an ancient writer is referring to, especially if the writer mentions only the name of the plant and does not describe it. Some plants, which have had the same names since antiquity, can be identified without difficulty.
>
> (Jashemski 1999: 19)

Jashemski developed new methods to determine which species grew in ancient gardens. For instance she created a method to study plant cavities (the cavities left by the roots of trees) by making plaster casts. The size of cavities, and the way they are spaced (whether there is a pattern or not) give indications on the type of plants they contained: vine, fruit trees, shrubs, etc.[53] Garden archaeologists deal with various forms of plant remains. Some are macroscopic (visible to the naked eye) and include carbonised and waterlogged remains. Carbonised fruits and vegetables have been found at various sites destroyed by fire and at Pompeii and Herculaneum (see Map 4). We could mention, among others, a date seed with a portion of the edible fruit found at Pompeii (see Jashemski 1974: 401); and burnt barrels containing pomegranates found at Vindonissa in Switzerland (see Map 3), dating to the first century CE (see Jacomet *et al.* 2002). Carbonisation could also be deliberate, as in the case of cereals exposed to fire. Waterlogged remains include wood samples from sunk ships, wells or cesspits. Both carbonisation and waterlogging preserve dry parts of plants better than soft tissues. As a result, most macroscopic remains are seeds, nutshells and fruit stones (see Bakels and Jacomet 2003). Other plant remains are microscopic (they can only be seen properly with the help of a microscope). Most important among microscopic remains are pollens. Pollen remains inform us about planting patterns and transplantation of plants (on such patterns, see Zohary *et al.* 2013).

Beside plant remains, garden archaeology studies all remains linked to gardens: water systems; garden buildings; planting trenches; planting pots (*ollae perforatae*), etc. Garden archaeology has now spread to all the regions of the

Roman Empire, and especially to gardens in modern Tunisia (see Jashemski *et al.* 1995; Jashemski 1996) and Britain (see Cunliffe 1971, 1981). On the other hand, very little has been done at classical Greek sites (see Megaloudi 2005; Megaloudi *et al.* 2007).

Garden archaeology allows to deepen knowledge of sites already well-known. For instance, Hadrian's villa has been studied by classical archaeologists and architects for centuries, but the gardens that were such an important aspect of the villa have been given scant attention until Jashemski and her students examined them (Jashemski *et al.* 1992). Another example would be the study (started in 1996, by the Swedish Institute) of the gardens of the Villa of Livia at Prima Porta, which had been hitherto neglected (see Klynne and Liljenstolpe 2000). But as Kathryn Gleason wrote, the plants of the garden remain its most elusive part:

> The species of plants and animals and the activities that animated garden life elude the archaeologist at most sites, although in certain conditions of preservation results can be surprising. Texts and art are needed to animate the remains and artifacts recovered from most garden surfaces.
>
> (Gleason 2010: 9)

Confronting texts and material remains, however, is a difficult enterprise. Texts follow complex conventions and agendas and should not be taken at face value. Ancient poetical descriptions of gardens, for instance, will often be filled with metaphors and allusions, and will therefore correspond to no reality found through archaeology.

1.2.4 Environmental studies

Our final type of source are the studies of environmental historians. These scholars often work on similar material to archaeologists (and collaborate with them), but take a long-term approach and work on large geographical areas. They ask whether the climate and environment of the Mediterranean changed significantly over the last 3,000 years (on the current vegetation of the Aegean world, see Davis *et al.* 1965–1988; Polunin 1980). The debates in this field are, to say the least, heated. Some scholars, most prominently Curtis Runnels (1995) and J. Donald Hughes (1983, 2011; Hughes and Thirgood 1982) argue that many areas of the Mediterranean were in the past covered with forest.[54] Because of man's overexploitation (through shipbuilding; fuel use; overgrazing; and deforestation to create arable soil), these forests started to disappear, leading to erosion and desertification. The forest gave place to the maquis. The ancients attempted to develop strategies to slow down this process, such as delineating sacred spaces (on which see Dillon 1997), but environmental degradation proved inexorable and led to the demise of the ancient empires:

> Forests provided the major material for construction and almost the only fuel source of the classical world, and depletion of this source precipitated a number of crises. As forests retreated with land clearance, wood decreased in availability and increased in price, contributing to the ruinous inflation that plagued late-antiquity. Competition for forest resources ignited military conflicts, which themselves created demands for timber. Erosion weakened the economic base of the predominantly agrarian societies, contributing to a population decline that made it ever more difficult for Greco-Roman civilisation to resist the incursions of barbarians from beyond the frontiers. In the more arid regions, forests that formerly moderated the climate and equalised the water supply were stripped away, permitting the desert to advance.
>
> (Hughes and Thirgood 1982: 60)

Such conclusions were reached through scientific analysis (for instance of charcoal remains or of pollen) and textual analysis. While the scientific analysis may be sound, the textual analysis is sometimes rather simplistic. Hughes tends to take the ancient texts at face value. This is particularly problematic when one of his key texts is Plato's description of a mythical Attica in the Critias (111c).

At the other end of the spectrum, Richard Grove and Oliver Rackham (2003) are very sceptical of environmental change. They also happen to be much more careful with ancient texts, which may lead the ancient historian to agree with their conclusions, perhaps unjustly.[55] They identify a 'Ruined landscape myth' with a very long – and prestigious – history:

> Renaissance poets and Baroque painters encouraged the belief that the actions of Antiquity took place in lands not too unlike the lush riversides of Normandy or the dramatic wooded badlands of the Papal States. Virgil, their inspirer, had not distinguished harsh Greece from idyllic Italy. When travellers reached the drier and remoter parts of the Mediterranean and compared what they saw with what they expected, they inferred that the landscape had gone to the bad since Classical times.
>
> (Grove and Rackham 2003: 8)

This, in their opinion, is no more than a myth. To expect the Mediterranean to be wooded with lofty trees is to apply North-American and North-European ideas of the ideal landscape onto an entirely different set of environments, for the Mediterranean is extremely diverse from an environmental point of view (see in particular 2003: 45 and 57). Grove and Rackham argue that Mediterranean vegetation was – and is – resilient, and that those who exploited the land (before the twentieth century) were careful to preserve its riches. They observe (2003: 169) that the Greek vegetation described by the traveller Pausanias in the second century CE was not very different from that observed by his nineteenth-century translator, Frazer. Peregrine Horden and Nicholas Purcell display the same scepticism towards the theory of extensive deforestation in their book *The Corrupting Sea* (2000: 330).

In order for definitive conclusions to be reached on this question, classicists, archaeologists, botanists and environmental historians need to join forces in multi-disciplinary themes. We note, however, that the latest – excellent – study on the notion of landscape in Roman literature (Spencer 2010) does not mention the works of Grove and Rackham.

Notes

1 Other mentions are to be found in Basil of Caesarea, *Homilies on the Six Days of Creation* 9.3, line 80 (*empeiria botanikē*); scholia in Homer's *Iliad* 11.846 (listed with *iatrikē*, the medical art); Tzetzes, *Chiliades* 6.94.963 (listed with *iatrikē*). On the rich preface to the *Materia Medica*, see Scarborough and Nutton (1982).
2 A similar division is to be found in the 'zoology and botany' article of the Neue Pauly (Hünermörder 2010), where material is divided into: 1) pure botany; 2) agricultural botany; 3) medicinal and pharmaceutical botany; and 4) astrological and magical botany.
3 See also Reed 1942: chapter 3 on antiquity; Mägdefrau 1973: 4–11; Morton 1981: 19–81; Segura Manguía and Torres Ripa 2009: 52–61. For a different type of approach to the history of ancient botany, see Baumann 1993.
4 Mycenean is an early form of Greek written in a syllabic system rather than the alphabetical system with which we are familiar.
5 For introductions to the Hippocratic Corpus, see Jouanna 1999; Craik 2014. For an introduction to ancient medicine more generally, see Nutton 2013.
6 Potter's edition for the Loeb collection has a different chapter numbering from that used by Littré and Lonie 1981. We have chosen to give Littré's numbering.
7 See e.g. *Humours* 11 (Loeb 4.82 Jones = 5.490 Littré): analogy between soil that nourishes plants and stomach of animals.
8 On identifications and studies of plants listed in the Hippocratic Corpus, see e.g. Dierbach 1824 (on all plants); Raudnitz 1843 (on all plants); Desautels and Girard 1983 (on all plants); Maloney 1989 (on plants in the treatises Epidemics); Girard 1990 (on hellebore); Moisan 1990a (on all plants); Moisan 1990b (on narcotic plants); Aliotta *et al.* 2003 (on all plants).
9 See also Hesychius, *Life of Aristotle* line 117.
10 Note however that *On Plants* is attributed to Aristotle in the Loeb Classical Library volume of Aristotle's minor works (Hett 1931).
11 Aristotle, *Meteorology* 1.1, 339a7; *On Sense and Sensible Objects* 4, 442b26; *On Length and Shortness of Life* 6, 467b4; *Enquiry into Animals* 5.1, 539a20; *Generation of Animals* 1.1, 716a1; 1.23, 731a29; 5.3, 783b20.
12 Aristotle, *Fragments* 267–278 Rose.
13 See the introduction to Mayhew's Loeb edition for a summary of the debate (Mayhew 2011; Mayhew and Mirhady 2011).
14 The best edition of Theophrastus' *Enquiry into Plants* is that by Suzanne Amigues for the Collection des Universités de France: books one and two (1988a); books three and four (1989); books five and six (1993); books seven and eight (2003); book nine (2006). Her French translation is also reprinted in a single volume beautifully illustrated (2010). The edition by Sir Arthur Hort for the Loeb Classical Library is also useful, especially for English speakers (1916–1926). Note, however, that a passage is

missing, as Hort censured 'indecent' parts of the text; see Gemmil 1973a. *Causes of Plant Phaenomena* is available in a Loeb edition and English translation by Einarson and Link (1976–1990). Amigues has also edited and translated into French books 1 and 2 (2012). *On Odours* is available in the second volume of Hort's edition (with an English translation) and in a more recent edition by Eigler and Wöhrle 1993. The fragments are edited and discussed by Sharples 1995. See the commentary to fr. 124 for the issues linked to incorrect citations. On other possible works, see Sollenberger 1988. The bibliography on Theophrastus' botany is extensive. See in particular the studies of Amigues (several which are collected in her 2002 volume), with in particular Amigues 1994 and 1999a as handy introductions; Strömberg 1937; Wöhrle 1985.

15 See Wöhrle (1985: 22–46) for a summary of the contents of Theophrastus' botanical works.

16 Wellmann (1898a) suggested that Diocles was the first ancient author to write a herbal, and that his work was a source for book nine of the *Enquiry*. Bretzl suggests the author is a root-cutter (1903: 366, note 24). Singer notes that 'The ninth book of the Theophrastean *Historia plantarum* is on a lower level than the rest, and is derived from writings of the nature of herbals' (1927: 2). Stannard writes that 'Book IX is possibly spurious' (1969: 213).

17 See Sollenberger 1988; Wöhrle 1988; Einarson and Link 1990: 459–463; Amigues 2012: XIII–XVII.

18 The best editions of Nicander are those by Gow and Schofield 1953 and Jacques 2002 and 2007.

19 First edited by Grenfell and Hunt 1922: 116–118. See also Page 1941: 506–509 (select papyrus n. 124). On Egyptian plants in Greek texts, see Marganne 1992; Amigues 1995c.

20 This poem is edited by Heitsch 1964: 23–38. See Luccioni 2006.

21 There are many editions of the *Georgics*, and many more studies. The edition used here is the Loeb Classical Library edition by Fairclough (revised by Goold, 1999). The *Georgics* is one of the best-known Latin poems, and many studies are devoted to it. Among recent studies Kronenberg 2009 and Thibodeau 2011 are particularly relevant.

22 On these plants, see the notes by Martyn (professor of botany at Cambridge) and King 1755; Fée (French botanist) 1822; Sargeant 1920; Abbe 1965.

23 The best edition of this text is that by Drossaart Lulofs and Poortman 1989. They give five versions of the text. The Greek text found in Hett 1957 is not reliable. On the compilatory method of Nicolaus, see Moraux 1973: 487–514; Drossaart Lulofs and Poortman 1989: 7. Herzhoff (2006) argues that Nicolaus had access to Aristotle's lost *On Plants*.

24 For a general introduction to the agronomists, see Diederich 2007.

25 There are many editions of Varro. The edition we have used is that of the Loeb Classical Library: Hooper and Ash 1934.

26 The edition we have used is that for the Loeb Classical Library in three volumes: Ash 1941 (books 1–4) and Forster and Heffner 1954 (books 5–9, which are less relevant to our purpose) and 1955 (books 9–12 and *On Trees*).

27 Scholars have also studied book ten of Columella's *On Agriculture* from a poetical point of view. See e.g. Henderson 2002.

28 Unlike the text of the other agronomists, that of Palladius has not known many editions. The edition used here is that by Rodgers 1975. A recent English translation is available: Fitch 2013.

29 Gargilius Martialis also wrote a treatise *On Drugs from Vegetables and Fruits* (*Medicinae ex holeribus et pomis*) which is better known, but contains very little botanical information, see Maire 2002.

30 The best edition of the *Geoponika* remains that by Beck 1895. A recent translation of the *Geoponika* is available in English: Dalby 2011.
31 The best edition of Dioscorides remains that by Max Wellmann: 1907 (books one and two); 1906 (books three and four); 1914 (book five; fragments of Sextius Niger and Crateuas; pseudo-Dioscorides' *On Simples*). The best English translation is that by Beck 2005. A recent German translation is also available: Aufmesser 2002. The best general study of Dioscorides' work is Riddle 1985. Touwaide 1999a is also a handy introduction to the work of Dioscorides.
32 Those views were expressed in a personal correspondence between Riddle and Nutton/Scarborough and is reported in Scarborough and Nutton 1982: 214–215. See also Riddle 1985: 4.
33 See Barbaud 1994.
34 Kästner edited this text in 1896. His edition is based on three manuscripts only and is defective.
35 There are numerous editions of Plinys' *Natural History*. Here, we have mostly used the editions for the Loeb Classical Library: Rackham 1968 (books twelve to sixteen); Rackham 1950 (books seventeen to nineteen); Jones 1969 (books twenty to twenty-three) and Jones 1980 (books twenty-four to twenty-seven). For French readers, the editions in the Collection des Universités de France are excellent: book twelve (Ernout 1949); thirteen (Ernout 1956); fourteen (André 1958); fifteen (André 1960); sixteen (André 1962); seventeen (André 1964a); eighteen (Le Bonniec and Le Boeuffle 1972); nineteen (André 1964b); twenty (André 1965); twenty-one (André 1969); twenty-two (André 1970); twenty-three (André 1971); twenty-four (André 1972); twenty-five (André 1974); twenty-six (Ernout and Pépin 1957); twenty-seven (Ernout 1959).
36 On Pliny's botanical and pharmacological knowledge, see Fée 1833; Brosig 1883; André 1955; Stannard 1965; Morton 1986; Scarborough 1986. Older scholarship focused very much on recovering the lost sources of Pliny – a method called Quellenforschung. See in particular Wellmann 1924.
37 The best edition of book seventeen is that by André for Les Belles Lettres (1981).
38 We will also encounter some other pharmacological treatises by Galen in this book. These are *On Antidotes* (in two books); *Composition of Medicines according to Places* (in ten books); and *Composition of Medicines according to Types* (in seven books).
39 The best edition of this text is that by Helmreich 1923 (CMG 5.4.2). The Kühn edition of this text is to be found in volume 6, pp. 453–748. For reliable translations into English, see Grant 2000: 68–190; Powell 2003.
40 There is a very considerable literature on the pharmacological writings of Galen. See in particular Fabricius 1972; the articles collected in Debru 1997; and Vogt 2009.
41 Edition: Kühn vol. 14, pp. 62–157.
42 An English translation is available for the works of Paul of Aegina. The translation of the relevant section contains numerous useful notes: Adams 1847: 17–479.
43 The text is edited by Friedrich 1968. The fullest manuscript in Greek in MS T (Madrid, Biblioteca nacional 4631). Friedrich's edition gives the text of Manuscript T in one column and other versions, which sometimes bear striking differences, in further columns.
44 In MS T the letter is by Harpocration; in MS M (Montpellier, Bibliothèque municipale, section médecine 277), a Latin manuscript, the letter is by Thessalus.
45 On the identification of Thessalus and Harpocration, see e.g. Scarborough 1987: 27; Scott 1991.
46 See for instance the Greek treatise *On the Herbs of the Seven Planets* (*De septem herbis planetarum*), edited by Sangin 1936: 126–135 and the Greek treatise *On the Herbs of the Planets* attributed to Alexander the Great, edited by Weinstock 1953: 129–135.

47 The fullest edition of the Cyranides is that by Kaimakis 1976.
48 The best edition is that by Calderón Dorda *et al.* 2003. A recent English translation is available online: www.luc.edu/roman-emperors/Pseudo-P%20Revised.pdf, accessed August 2015.
49 These texts are edited by Howald and Sigerist 1927. See Maggiulli and Buffa Giolito 1996 on the issues of the 1927 edition, and the need for a new one. For an introduction to the anthology, see Jones 1999.
50 See McEnerney 1983.
51 For other examples of short botanical treatises from late antiquity and the early Middle Ages, see Thomson 1933 and 1955.
52 Pompeii and the other Vesuvian sites remain extremely important sites to archaeobotanists and environmental historians. See e.g. the essays edited by Jashemski and Meyer 2002; Ciarallo 2004.
53 See for instance Jashemski 1974. Kathryn Gleason has created a new technique using silicon rubber to create casts of smaller cavities (Dragon Skin TM) (2010: 12).
54 Thommen (2012: 37–41) is also a recent proponent of the deforestation theory.
55 Grove and Rackham criticise Hughes' use of ancient texts (see in particular 2003: 18 and 288). They re-assess the ancient literature in chapters 9 and 11.

2

ACQUIRING KNOWLEDGE OF PLANTS IN THE ANCIENT WORLD

2.1 Introduction

> My admiration of antiquity certainly grows when I consider the very treatment of this subject [i.e. the study of plants]; the larger the number of plants remaining to be described, the more one is led to honour the ancients for the care they took in their studies, and their benevolence in transmitting their findings.
>
> (Pliny, *Natural History* 27.1)

On multiple occasions in the *Natural History*, Pliny the Elder presents the study of plants as a field of knowledge that has a prestigious history, stretching from the mythical past to the current day and looking into the future, with numerous discoveries along the line – discoveries of new plants; discoveries relating to their properties; and discoveries of how to make them grow better or differently.[1] In the distant, mythical past, heroes and gods were responsible for such finds. For instance, Apollo, the first plant explorer, discovered the peony (*HN* 25.29).[2] In the historical past, poets, philosophers, travellers, kings and generals increased botanical knowledge (see examples below). In order to make such discoveries, Pliny's 'ancients'

> explored also inaccessible mountain tops, unknown deserts, and all the bowels of the earth, to discover the power of every single root, and the uses to which mere threads of herbs can be suited, and turning to healthy use those that even animals did not touch as fodder.
>
> (Pliny, *Natural History* 25.3)

They then generously shared their knowledge with the world, kindly putting it in writing, unlike moderns:

> We desire to hide and conceal what they [the ancients] took pains to discover, and to deprive human life even from those foreign goods. For those who have gained a scrap of knowledge hoard it jealously away

> from others, and through their teaching nobody, their learning is held in authority (*in auctoritatem scientiae*).
>
> (Pliny, *Natural History* 25.2)

There is little doubt that Pliny's dichotomy between the generous ancients and the secretive moderns is but a trope.[3] Pliny does mention relatively recent finds that have been shared with the world: for instance that of a vine producing a wine with a natural flavour of pitch, which 'was unknown in the period of Vergil, who died ninety years ago' (*HN* 14.18); or that of another vine at Viviers in the province of Narbonne that is immune to bad weather, discovered 'in the last seven years' (*HN* 14.43); or the recent discovery of *hiberis* by the physician Servilius Damocrates (fl. towards the end of the first century CE, see Vogt 2008) (*HN* 25.87). It is also clear that Pliny's insistence on discovery and sharing of knowledge is not innocent. In the encyclopaedist's mind, plant discoveries are nothing but Nature revealing itself in all her bountifulness:

> Now it is clear that the bountifulness of Nature is the work of the gods, or that it is at least divine, even when a man discovered [plants and their properties], and that the mother (*parentem*) of all things both generated them [the plants] and showed them to us.
>
> (Pliny, *Natural History* 27.2[4])

Thanks to the 'boundless grandeur of the Roman Peace' knowledge of this provident Nature, in Pliny's time, should have been possible on a scale hitherto unknown (*HN* 27.3): there was a clear link in the encyclopaedist's mind between the growth of empire and botanical discoveries.[5] However several factors collided with the favourable conditions afforded by the Roman Empire: first, chance plays an important role in botanical discoveries – nature may choose or not to reveal itself; second, men tend to forget about ancient finds; third, people were now more interested in luxurious items than in such simple things as plants:

> Formerly, it used to be an ambition to give one's name to plants, as we will show kings did. To discover a plant was considered a great achievement, and thus to assist human life, but now some will judge that my research in this field is worthless – so despised in comparison to luxuries are even those things that bring us health.
>
> (Pliny, *Natural History* 25.22)

It is therefore plain that Pliny's constant allusions to the prestigious history of botany and herbal healing are ideologically motivated. These references, however, raise the important question of how botanical knowledge was acquired, transmitted and retained in the ancient world – the question at the centre of this chapter. At the beginning of *On Agriculture* (1.1.11), Varro explains that he acquired his agricultural knowledge from three types of sources: his personal experience; what he had read; and what he had heard.[6] Dioscorides too, in the preface to the *Materia Medica*, lists these three sources of information:

> For I have, with the greatest of precision, got to know most of my subject through observation (*di'autopsias*), and got to understand thoroughly the rest through what is universally accepted in the written record (*ex historias*) and in making enquiries (*anakriseōs*) of the natives (*epichōriōn*) in each case/region.
>
> (Dioscorides, *Materia Medica*, preface 5)

Through these three means, Dioscorides has acquired experience (*empeiria*), unlike those of his predecessors who relied most heavily on the written word. Dioscorides, writing in the first century CE, had a fully-developed notion of *empeiria*, but Theophrastus much earlier also insisted on the centrality of a multi-faceted experience (see below for more detail).

The epilogue to the *Alphabet of Galen*, which is probably later than the main text, also mentions three sources of medico-botanical knowledge: memory (*memoria*), usage (*per nos ipsos probauimus et quae experti sumus*: 'those plants which we ourselves have tried and tested'), and ancient authorities (*apud antiquos auctores*) (*Alphabet*, epilogue, 380 Everett). However, in the chapters devoted to plants, the *Alphabet* mentions no author whatsoever.

The three categories – 'personal observation', 'oral sources' and 'written sources' – shall serve as structuring tools in this chapter. We use the categories in full awareness that there might be overlap between them, and that it can at times be difficult to determine in which category to place some sources. For instance, ancient botanists often travelled to observe vegetation (personal observation), but they did not always visit the exact places where plants grew, especially when this required climbing a high mountain in sweltering heat, relying instead on the accounts of local people (oral sources). Also, it is difficult to determine whether a source is oral or written – or even both – when our texts use phrase such as 'it is said', 'some think', etc. As research on orality and literacy has pointed out, there are complex interactions between oral and written traditions in the ancient world.[7]

Throughout this chapter we will encounter the actors of ancient botany – those who, beside the authors presented in Chapter 1, collected, handled, described and discussed plants. These range from the illiterate peasant to the mightiest king, and also include philosophers, gardeners, root-cutters, magicians and poets. This diversity leads us to the question of expertise. In our modern world, we give the label of 'expert in botany' to people who have a university degree in the topic or to people who, through years of experience, have gained significant knowledge of the field in all its technical aspects. The notions of expertise and authority were far more fluid in antiquity, and we will see that even Homer, who mostly mentioned plants in passing, was considered to be a plant expert.

Figure 2.1 Seven ancient authors active in the fields of botany and pharmacology, clockwise from top right: Dioscorides, Nicander, Rufus, Andreas, Apollonius Mys, Crateuas, Galen (centre). MS Vienna, Österreichische Nationalbibliothek medicus graecus 1 (512 CE), fol. 3v. Courtesy of Österreichische Nationalbibliothek, Vienna.

2.2 Personal observation

The most basic form of personal botanical experience was – and still is – to observe plants. All ancient botanical writers claim to have observed plants for themselves in their natural environment (*autopsia*). Some plants were common and could be observed almost anywhere in the Mediterranean basin, although it was sometimes useful to note special characteristics linked either to a particular locale or to a particular plant. Ancient writers also enjoyed pointing out marvellous and amazing botanical specimens. For instance, Theophrastus noted that pieces of armament were found within the trunk of an oleaster growing on the agora of Megara (see Map 1), when that tree was cut down, adding that there was now very little left to see of that tree (*HP* 5.2.4). The same author mentioned that a young plane tree growing in the Lyceum had roots 'spreading over 33 cubits' (*HP* 1.7.1). Interestingly, that mention of the Lyceum is unique in Theophrastus, who also makes scant reference to Athens in general (see Thanos 2005a). It is also important to note that both the Megara-tree and the Lyceum-tree are cultivated. Indeed, Theophrastus conceded that few people were experienced (*empeiroi*) in the field of wild trees, while the knowledge (*aisthēsis*, that is, knowledge through the senses) of cultivated ones was more common (*HP* 1.14.4).[8] Theophrastus, however, as any good Aristotelian philosopher would have done, warned against unmediated sensory observation: it must be confirmed by reasoning, *logos* (see e.g. *CP* 1.1.1). When not taking *logos* into account, one runs the risk of generalising the particular – believing that what happens to plants in one region occurs everywhere (see *CP* 3.2.3–4).

Dioscorides also insisted on observing plants with one's own eyes (*autopsia*), not just once, but at various stages of their developments. Failure to do so may lead to various errors:

> He who wants to gain experience (*empeirian*) in these matters must be present when they are newly sprouted from the ground, as well as when they are in their prime and past their prime. For neither he who has encountered a plant only at the sprouting stage can recognise it when in its prime, nor he who has seen it at its prime can identify it when sprouting. For because of changes in the form of the leaves and the size of the stems, flowers and fruits, and because of some other specific characteristics, people who did not make their observation (*thean*) in this manner made great mistakes concerning some plants.
>
> (Dioscorides, *Materia Medica*, preface 7)

Dioscorides then proceeded to condemn some writers (probably the 'followers of Asclepiades' criticised a bit earlier in the preface) for their lack of observation. In the body of his *Materia Medica*, Dioscorides sometimes gave examples of mistakes committed through lack of observation. For instance, he wrote that people have supposed colt's foot (Greek: *bēchion*, *Tussilago farfara* L.) to be flowerless because they did not observe it in the spring, before it had lost its pallid flower (*MM* 3.112).

Galen, who like Theophrastus insisted on using observation together with reasoning (see van der Eijk 1997), also enjoined his readers to follow the growth of medicinal plants, unlike the perfume-sellers who only knew them in their dried state:

> For, when you can recognize them from their birth to their maturity, you will be able to find them in many places of the earth. Thus I have found in many regions of Italy plants which some could not recognise when sprouting and growing, as they had only seen them in their dried state. There is not a single perfume-seller (*myropōlōn*) who does not know the plants imported from Crete, as well as their fruits, but they do not know that many of them grow in the vicinity of Rome.
>
> (Galen, *On Antidotes* 1.5, 14.30–31 Kühn)

In order to observe plants at all stages and to control their provenance, some people involved in healing planted their own gardens. Such was the case of Antonius Castor, an esteemed pharmacologist (first century BCE–first century CE, see Keyser 2008a), in whose garden Pliny observed a large array of plants:

> At least it was possible for me to observe most plants, with only a few exceptions, thanks to the skill (*scientia*) of Antonius Castor, who was the highest authority of our time in this art [botany]. I used to visit his little garden (*hortulus*) in which he grew many plants, even when he had grown older than one hundred, having experienced no bodily ailment, and in spite of his age, no impairment to his memory or vigour.
>
> (Pliny, *Natural History* 25.9)

Unfortunately, we know nothing else about Castor's garden, yet some have not hesitated to qualify it as a 'botanical garden'. Similarly, people have interpreted Theophrastus' garden, listed in the philosopher's will (see Chapter 1), as a 'botanical garden'.[9] This phrase implies a certain level of systematism and a defined research purpose that might not have been present in those two ancient gardens. It is probably best to see Castor's garden as a health garden, and that of Theophrastus as a philosophical garden, in which learned men enjoyed walking, debating and indeed observing plants. That being said, some conditions were present in the Hellenistic and Roman periods for the birth of the botanic garden, when kings, generals and other prominent figures attempted to transplant foreign species to their estates. In fact royal gardens and parks (*paradeisoi*), with their rich collections of plants, may have been an important source of information for any botanical writer (see Chapter 6).

Not all plants could be successfully grown in gardens, however, and some changed in the process of transplantation from the wild to a cultivated garden (becoming barren for instance). The ancients, therefore, saw travelling in order to observe plants in their native environment as a necessity. The extent of Theophrastus' voyages is a matter of debate, as he does not mention the act of travelling. It is reasonable to assume that he journeyed extensively through mainland

Greece and the islands of the Aegean to examine plants in their natural habitat: the vegetation of Orchomenus (Boeotia, see Map 1); Lesbos (in particular Pyrrha and its mountains, see Map 1); Macedonia (to which Theophrastus may have had connection through Aristotle, see Map 1) and Chalcis (Euboea, see Map 1) in particular get extensive treatment. Slightly further afield, but still in the Greek-speaking world, Theophrastus has detailed knowledge of the vegetation of Crete (see Map 1), of Mount Ida of Troad (which he may have explored while on a visit to Aristotle in Assos, modern Turkey, see Map 1), and of Cyrene (see Map 2), although he may not have observed Cyrene's most famous product – silphium – in its natural habitat (*HP* 4.3.1).[10] Whether Theophrastus travelled to those regions newly conquered by Alexander the Great (see Map 2) is a more complicated matter. He mentions the Great Macedonian king on several occasions (*HP* 4.4.1, 4.4.5, 4.7.3), and often refers to Babylonia, Syria and Egypt, and in a way that could lead us to believe he had been there himself. However, similarities in his accounts with those found in the historical works of Arrian (*c.*86–160 CE, *Anabasis* and *Indica*) indicate that both the philosopher and historian used the same sources: the reports of those who accompanied Alexander's expeditions. It is, however, impossible to determine whether Theophrastus discussed these matters with people such as Androsthenes (FgrH 711, see Dognini 2008), Aristobulus (FgrH 139, see Kaplan 2008a), Nearchus (FgrH 133, see Kaplan 2008b) and Onesicritus (FgrH 134, see Kaplan 2008c), or whether he read their works (all of which are now lost, save for fragments).[11] It should also be noted that of these four authors, Theophrastus only names one – Androsthenes – on one occasion (*CP* 2.5.5).

While Theophrastus did not mention the act of travelling, Dioscorides himself reports that he travelled extensively thanks to his 'soldier-like life' (see Chapter 1). The extent of Dioscorides' travels is, however, also a matter for debate. John Riddle (1985: chapter 1) argued that the pharmacologist saw most places mentioned in the *Materia Medica*, visiting locations as far west as Spain and as far south as Egypt, and including some localities outside the Roman Empire. Indeed, Dioscorides gives very good information on Petra, the capital of the Nabataean Arabs (see Map 3), which only became Roman in 166 CE. Alison Denham and Midge Whitelegg (2014), on the other hand, noted that Dioscorides only uses the first person once outside his preface (a reference to the Mountains of the Vestini, in central Italy, *MM* 2.70, see Map 4), and concluded that he did not travel much at all. However, we do not find this argument compelling, since if we followed it to its logical conclusion, we would have to argue that Dioscorides did not observe plants in his native Cilicia, of which he talks regularly, but never in the first person (see e.g. *MM* 3.97).

Galen, for his part, extensively reported his travels in his writings, many of which have an autobiographical element. He was familiar with his native Cilicia, various islands of the Aegean, the region around Alexandria (see Maps 2 and 3), Syria, Rome, etc. At times he makes comments about the plants he saw during his travels, as for instance:

> In Alexandria, I saw the sycamore tree, with fruit resembling a small white fig ... I also saw in Alexandria the persea, which is one of the large trees there. They say (*historousi*) that its fruit is so harmful in Persia that it kills those who eat it, but that when imported into in Egypt it becomes edible.
>
> (Galen, *Properties of Foodstuffs* 2.35 and 36, 6.617 Kühn[12])

Note how Galen in the same paragraph combines personal observation and oral information, which he does not entirely trust, prefacing it with the words 'they say' (see pp. 41 and 46 for more detail). In general Galen's botanical *autopsia* was not as extensive as that of Dioscorides, to whom he often refers. On one occasion the physician acknowledges that he has no experience of the plant he is discussing (a plant named *asclepias*), and instead relies on the authority of Dioscorides (*Simples* 6.1.64, 11.840 Kühn; see Dioscorides, *MM* 3.92). On the other hand, Galen once claimed superior knowledge of a plant: unlike Dioscorides, he knew that fresh Damask plums (and not only dried ones) have a staunching power (*Simples* 6.4.1, 11.861 Kühn; see Dioscorides, *MM* 1.121).

Pliny too observed plants while serving in the army (campaigns in Upper and Lower Germania, see Map 3) and as a procurator in the provinces of Africa and *Hispania Tarraconensis* (see Map 3). For instance, he reported having observed Daphnis' cassia 'on the extreme edge of our empire, where the Rhine washes our frontier, planted among beehives' (*HN* 12.98); or that in Lacetania (part of Spain closest to Italy) he learned of the discovery of a plant called *dracunculus* (*HN* 25.18). These claims show Pliny as an authority in the field of botany, not just a bookish scholar (see Stannard 1965). Columella also appears to have travelled, as he reports having seen plants harvested a few months after sowing in parts of Cilicia and Syria (*RR* 2.10.18). The interlocutors in Varro's agricultural dialogue also tell of the flora they saw – or did not see – on their travels. For instance, Scrofa recounts how, when he was commanding the army in Transalpine Gaul near the Rhine (see Map 3), he saw places where neither vine nor olive grew (*RR* 1.7.8).

Some ancient authors, not content with mere observation, had a direct involvement with plants. Thus, Dioscorides, Galen and other physicians/healers handled plants for their pharmacological preparations; and Roman agronomists took pride in their personal growing and caring for plants. Columella in particular insists on the importance of experience (*experientia*) and practice (*usus*), which he calls the 'master of the art' (*RR* 1.1.16, see p. 51 for more detail). Much of his experience he gained in the company of his paternal uncle, Marcus Columella (*RR* 2.15.4, 5.5.15). This great practical knowledge led him to invent new cultivation methods. For instance, he tells us that he devised a way to combine trees of different natures, such as figs and olives, through grafting (*RR* 5.11.12–15; see Pliny, *HN* 17.137). Whether he himself invented the method, or whether it was designed on his estate, makes relatively little difference: this discovery was made under his patronage and was therefore his to claim.

Palladius too often wrote in the first person about his farming experiments, using his personal experience on his estates near Rome and in Sardinia as a way either to counter or strengthen an argument he read in his written sources, as in the following example:

> Many [authors] give different dates for the sowing of quinces. But I have found out through experience (*mihi usu*) that, in Italy around Rome, quince plants rooted in February or at the beginning of March in prepared ground took so beautifully.
>
> (Palladius, *Work of Agriculture* 3.25.20[13])

Palladius regularly made such claims in relation to fruit trees, and in particular in relation to grafting methods (see Chapter 5).

While ancient botanical writers insisted on seeing plants for themselves, and at times on personal involvement with plants, they recognised that there must be a limit to their personally acquired knowledge. In some cases, this limit could only be acknowledged, in others one could rely on data provided by others, either orally or in writing. We turn to the former.

2.3 Things heard

In this section, we will discuss how Greek and Roman botanical writers discussed plant matters with 'natives' and with members of 'professional' groups – we use quotation marks as both terms are problematic. A discussion of the role of root-cutters (Greek: *rhizotomoi*) will lead us to consider the boundary between magical and non-magical practices in ancient botany. We will then turn to those stories that are introduced by phrases such as 'they say' or 'it is said' in the ancient texts.

As mentioned above, Dioscorides listed the 'natives' as one of his sources (*MM*, preface 5). The Greek word is *epichōrioi*, literally 'those who live in the region, the inhabitants' – it has none of the negative tones that can be attached to the English word 'native'. In the body of his *Materia Medica*, Dioscorides refers on numerous occasions to these natives, sometimes giving detail which one could term ethnobotanical, as in the example on how the Libyans collected spurge:

> Spurge: it is a fennel-like Libyan plant[14] growing in Autololia near Mauretania, full of the sharpest juice, which, on account of the strength of its burning, the men in that region collect with anxiety. Therefore, they wash sheep's entrails and tie them to the plant, then they cut off the stalk with javelins from far away. Immediately a great amount of sap pours into the entrails, as if from some sort of vessel.
>
> (Dioscorides, *Materia Medica* 3.82.1)

In practising a form of ethnobotany, Dioscorides was following in a longstanding tradition established by Theophrastus and in which all our main sources participated: Pliny, Galen, Columella, etc.

Theophrastus often mentioned the people of the Ida, the Arcadians, the Macedonians, the Boeotians and the Thessalians as sources of information (see Map 1 for the location of these peoples).[15] These peoples had their own names and modes of classification for plants. For instance, Theophrastus informs us that the Arcadians called the tree elsewhere known as *itea* (willow) *helikē* (*HP* 3.13.7); and that the people of the Ida distinguished five 'types' (*genē*, see Chapter 3) of oak, while the Macedonians only had four (*HP* 3.8 2–7). Peoples were also an important source of information on plant use. For instance, according to the philosopher, the Macedonians only used the 'male' variety of pine to produce pitch, while of the female pine, they did not even employ a part of the root (*HP* 9.2.3, see Chapter 5 for male and female plants).

It is worth considering which peoples were the objects of such ethnobotanical enquiries. Theophrastus never referred to his own people, the Eresians or the Lesbians, in the *Enquiry into Plants*. This silence is probably the result of some social convention that prevented him from discussing people too close to himself. Dioscorides, by contrast, referred to his kinsmen, the Cilicians as *epichōrioi* – his ethnobotanical research extended to his own region. Galen too employed the word *epichōrioi* in relation to his own people, the Pergamene (*On the Thinning Diet* 12.102, CMG 5.4.2, p. 449 Kalbfleisch), as well as in relation to his adoptive people, the Romans (*Composition of Medicines according to Types* 1.15, 13.428 Kühn). However, when talking about Pergamum, he preferred the phrase 'among us' (*par'hēmin*), as in this example: 'Men among us call the fruit of the bramble *bratina*' (*Foodstuffs* 2.13, 6.589 Kühn).[16]

Two peoples were especially renowned for their knowledge of plants in antiquity: the Thessalians and the Marsi. The Thessalians specialised in the preparation of philtres and love potions, and some Thessalian women were described as witches (Greek: *pharmakeis*), descendants of Medea (see Phillips 2002). Dioscorides reports how 'Thessalian women' used the roots of orchid and vetch as love potions and aphrodisiacs (*MM* 3.126 and 4.131). The Marsi were a people from central Italy renowned for their cures of serpent bites by charms and herbal remedies. According to Pliny, they were the descendants of the Italian Circe, through her son (*HN* 25.11). For that reason, the tragic poet Aeschylus had called Italy 'the Tyrrhenian race, a nation producing drugs' (fr. 2 West, on which see Theophrastus, *HP* 9.15.1). Pliny also reported that a plant called *consiligo* was discovered among the Marsi in his day (*HN* 25.86). Palladius added that this plant grows in the Marsian Mountains, and that it should be 'dug up with the left hand before sunrise, for it is believed to be more potent when so gathered', thus alluding to the ritual practices observed by the Marsi (*OA* 14.4.5).

At home as on their travels, ancient botanical writers made enquiries of people who encountered plants on a daily basis in their job: gardeners; farmers; vinelayers; pruners; beekeepers; shepherds or herdsmen; people involved in the handling of flowers (wreath-makers); people involved in the handling of wood (woodcutters; joiners/carpenters; torch-makers; master builders); in the drug business (root-cutters; root-sellers; perfume-makers; physicians); and merchants.[17]

Botanical writers also encountered people involved more incidentally in the handling of plants: priests (who burn incenses or use other plants, see e.g. Dioscorides, *MM* 3.23.5); smiths (see e.g. Theophrastus, *HP* 5.9.3); tanners and dyers (who use some plants as dyes, see e.g. Dioscorides, *MM* 1.108.1 and 2.184); the wool-cleaners (who use soapwort to wash wool, see e.g. Dioscorides, *MM* 2.163); divers (who observe the submarine vegetation, see e.g. Theophrastus, *HP* 4.6.9); and soothsayers (who see omens in various plant characteristics, see e.g. Theophrastus, *HP* 2.3.2). These 'professional' groups are mentioned most often in relation to plant usage, but their particular nomenclature or classifications are also noted. For instance, Dioscorides notes that unguent-makers give the name *pelekinos* to the plant he calls *hēduasaron* (*MM* 3.130).

It is important to note that these people appear to have willingly shared their knowledge and acted with generosity and without secrecy, unlike the 'moderns' mentioned by Pliny at the beginning of the chapter. Pliny, however, noted that some information on plants could be lost because people who knew them were illiterate: 'But the reason why more [herbs] are not known is because those who experience them (*experiuntur*) are illiterate (*litterarumque ignari*) folk people (*agrestes*), who are alone in living among them (*inter illas vivant*)' (*HN* 25.16). Pliny is being rather unfair here: while many local plant experts might have been illiterate, it is dangerous to assume they all were. Pliny may have his reasons for thus conflating various categories, but it is important that the modern scholar does not follow him. As stressed by Rosalind Thomas, it is crucial not to equate orality with folklore or with 'popular' culture (1992: 7).

At times, our authors 'promoted' one or several members of a professional group by identifying them as 'experienced' or 'careful'. Thus Theophrastus wrote that experienced gardeners (*empeirous tōn kēpourōn*) recognised that vegetables were often hungry (*HP* 7.2.5); and Palladius (*OA* 3.17.6) reported a grafting method he obtained from a meticulous farmer (*diligens agricola*). In this way, these authors vouched for the reliability of their sources, while keeping them in rather convenient semi-anonymity. Indeed, they rarely gave details as to where, when, or even whether, the exchange happened, using instead vague phrases such as 'the beekeepers say' (*HP* 6.2.3).

The professional groups most often mentioned are those involved in the trade of pharmacologically active plants: the root-cutters (Greek: *rhizotomoi*), the drug-sellers (Greek: *pharmakopōlai*; Latin: *pharmacopolae*), the perfumers (Greek: *myrepsoi*, *myropōlai*; Latin: *myropolae*, *thurarii*) and the herbalists (Greek: *botanikoi*).[18] We discuss these among 'oral sources', but we need to make clear that some of them were literate and even wrote very famous botanical works – it is simply not safe to assume that these 'plant handlers' were illiterate or somewhat inferior to physicians. Several 'plant handlers' were wealthy enough to purchase funerary inscriptions (see e.g. *CIL* VI 5639; on which

Korpela 1995). Some *botanikoi*, according to Galen, were in the service of the Emperor on the island of Crete (Galen, *Antidotes* 1.3, 14.9 Kühn). The link between royalty and herbalists does not stop there. For example, Antonius the *rhizotomos*, whom Galen qualifies as 'having much pharmacological experience' (*Composition of Medicines according to Places* 2.2, 12.580 Kühn), may be the same person as Antonius Musa, the famous physician to the emperor Augustus (see Chapter 1). In the second century BCE, Crateuas the *rhizotomos* (see Figure 2.1), one of the most diligent writers on botany according to Dioscorides (*MM*, preface 1, see Jacques 2008c for references, see Chapter 4 for Crateuas and plant illustrations), was a correspondent of Mithradates VI, King of Pontus, and dedicated a plant to him (*HN* 25.62). Scholars have often suggested that Crateuas lived at the Pontic Court, and that he assisted Mithradates in developing his famous antidote against all poisons (Mithridatium).[19] Unfortunately, nothing can prove this hypothesis.

'Crateuas' might have been a common name among 'dynasties' of herbalists – transmission of knowledge in the ancient world usually happened within families. Indeed, the comic poet Alexis (fourth-century BCE) composed a play entitled 'Crateuas the *pharmakopōlēs*', unfortunately only known through a handful of fragments (fragments 115–120 Kassel and Austin). A letter from Hippocrates to Crateuas, the 'excellent *rhizotomos*' concerning the madness of the philosopher Democritus is also preserved (*Letter* 16, 70–73 Smith 1990 = 9.342–348 Littré, see Chapter 1). This letter is certainly pseudepigraphic, but it remains possible that yet another Crateuas was a close contemporary of Hippocrates. According to the letter, Crateuas inherited his name and his trade from his grandfather.

Theophrastus has a long vignette on the activity of six *pharmakopōlai* in book nine of the *Enquiry*: the very skilful Thrasyas of Mantinea; his student Alexias, 'also experienced in the art of medicine'; the 'famous' Eudemus; Eunomus of Chios: Aristophilus of Plataea; and 'the Indian' (*HP* 9.16.8, 9.17.1–3 and 9.18). Thrasyas, Alexias and Eudemus are mentioned in relation to dangerous drugs: one that causes 'an easy and unpainful death' made of hemlock and opium-poppy (*HP* 9.16.8); and hellebore, which they attempted – with varying levels of success – to consume without being purged (*HP* 9.17.1–3). Aristophilus and 'the Indian' are named in a paragraph on plants that increase sexual potency (*HP* 9.18.4 and 9).

While ancient authors such as Theophrastus, Pliny and Dioscorides acknowledged 'herb handlers' as their sources, they also regarded them with a certain amount of suspicion. As already mentioned, Galen criticised perfume-makers, who only knew plants in their dried state. Pliny went even further in his accusations: he criticised the herbalists (*herbariori*) for their dishonesty (*scelus*):

> They keep a part of it [an iris] and of some other plants, such as plantain. If they consider their pay too little and look for further work, then, they bury in the same place the part they have kept, I believe so that the ailments they have cured would break out again.
>
> (Pliny, *Natural History* 21.144)

Both the encyclopaedist and Dioscorides noted on numerous occasions how plants, and in particular exotic plants, were adulterated, most probably by unscrupulous plant handlers (see e.g. Dioscorides, *MM* 3.22.3 on the adulteration of aloe).

It was the rituals involved in herb gathering, however, that attracted most criticism from our sources. Theophrastus criticised the *rhizotomoi* and *pharmakopōlai* for their methods of plant gathering that were often 'over-dramatized' (*epitragōdountes*). While it was at times legitimate to recommend gathering plants facing away from the wind; or covered in oil; or at a certain time of the day, other recommendations were, in his eyes, not acceptable:

> But the following notions seem ludicrous and far-fetched, for instance, they order that peony, which some call *glukusidē*, be dug up at night. For, if a man should do it during the day and be caught by a woodpecker while he is collecting the fruit, there is great danger for his eyes; and if he cuts the root then, his anus will prolapse.
>
> (Theophrastus, *Enquiry into Plants* 9.8.6)

Slightly further, he discusses whether it is necessary to pray when gathering plants:

> To pray while cutting herbs is perhaps nothing extraordinary, but it is when they add one or the other recommendation, as in the case of the type of all-heal called 'of Asclepius'. For then one should throw on the ground in its place offerings of all sorts of fruits and a honey-cake.
>
> (Theophrastus, *Enquiry into Plants* 9.8.7)

Galen for his part condemned charms that were pronounced over plants. He did so in his criticism of the 'grammarian' Pamphilus (see Keyser 2008c for references), who wrote a (now lost) treatise on simple medicines ordered alphabetically:

> He was diverted towards some old wives' stories and some silly Egyptian magical tricks together with some charms, which they pronounce over the plant when they pick it up. And he used them for amulets and other magical tricks (*manganeias*), which not only are superfluous and not part of the medical art (*iatrikēs technēs*), but altogether false.
>
> (Galen, *Simples* 6, preface, 11.792 Kühn[20])

A little further, Galen accused Pamphilus of using books attributed to Hermes the Egyptian which described the thirty-six plants of the horoscope, material which the physician found 'laughable'. The Pergamene then concluded: 'but Pamphilus, like many others, may have had the leisure to record useless stories. We, for our part, think we have wasted too much time recording them' (*Simples*

6, preface, 11.798 Kühn). Examples of 'Egyptian charms' are preserved in the Greek magical papyri, which contain among other things, a long prayer to be pronounced when gathering herbs (PGM 4.2967–3006).[21] This prayer mentions various gods (Cronos, Hera, Ammon, Isis, Zeus and Helios) and is accompanied by a complex ritual.

While Theophrastus and Galen were critical of what they considered to be superstitious or magical practices, other authors had a slightly different attitude. Cato included in his writings a prayer for the thinning of a grove (*RR* 139); Columella recorded a prayer for the sowing of radishes (*RR* 11.3.62); Dioscorides mentioned at least one amulet without any sign of criticism (*MM* 3.11); and pseudo-Apuleius' *Herbarius* includes complex instructions on how to draw circles with a sword around the mandrake (see Figure 2.2) before uprooting it, at night, with the help of a dog (chapter 131, p. 222 Howald and Sigerist).[22] Ancient authors, then, had diverse notions of the boundary between acceptable and unacceptable practices in the gathering and application of plants. Even those most critical of magical practices did not reject them outright. Instead they introduced their reports with phrases such as 'they say'; 'it is said' or 'some say'; or 'they think'. For instance, Theophrastus writes that 'it is said that cumin should be sown while cursing and insulting it' (*HP* 7.3.3). We would argue that these phrases are not outright rejections, but rather a way to establish a distance from, or to suspend judgement on, a practice that is still worth noting. Indeed, the same phrases are often employed when a fact could not be verified. For instance, Dioscorides writes that 'it is said (*historeitai*) that there is another species of *kuperos* that grows in India, similar to ginger' (*MM* 1.5). Since Dioscorides probably did not travel to India, he could not verify this fact and therefore established a distance between himself and the claim, even though it might have been valid.

It is important to stress that phrases such as 'they say' do not necessarily refer to oral material. Ancient writers sometimes used them in relation to written material; reading out loud was the norm in the ancient world. Nevertheless, these phrases are employed in relation to two forms of knowledge that most probably originated in oral discourse: proverbs and legends/stories that surrounded plants, to which we now turn. Advice on cultivation of plants was, in the ancient world as it still is today, transmitted in the form of short, aphoristic proverbs. For instance, Theophrastus writes that 'They say, speaking in proverbs (*paroimiazomenoi*) that "it is the year that bears and not the field"' (*HP* 8.7.6); Pliny reports that 'Syria takes great pains over its gardens, hence the Greek proverb (*proverbium*) "numerous are the Syrian vegetables"' (*HN* 20.33); and the sixth chapter of Palladius' first book is a collection of such maxims.

Botanical authors could at times be critical of proverbial knowledge, as in the following example:

> For a rich and good soil can bear more than one that is sandy and light. But there is a certain saying (*legetai tis logos*) whereby the same soil can at times take more [seed] and at others less. And when it takes more

Figure 2.2 A female mandrake being pulled up by a dog. Illustration to pseudo-Apuleius' *Herbarius*. MS London, Wellcome 573 (mid-thirteenth century), fol. 53v. Courtesy of the Wellcome Library, London.

> it is considered as a bad omen (*oiōnizontai*), for they say (*phasi*) at once that the earth is hungry. However this is perhaps a rather silly saying.
>
> (Theophrastus, *Enquiry into Plants* 8.6.2)

Yet again, that knowledge was worth engaging with, as it was an important aspect of Greek culture. In the same way, it was worth engaging with stories surrounding exotic plants. For instance Theophrastus reports the following story about *kinnamōmon,* a plant whose identity and origin was rather debated in antiquity:

> There is a story (*muthos*) told on the subject of *kinnamōmon*. For they say (*phasi*) that it grows in ravines full of snakes whose bite is fatal. Against these, men protect their hands and feet before descending in the ravines and gathering the plant. Then, when they have brought it back, they divide it in three parts and allot one to the god Helios. And that allotted to Helios, they leave there, and as soon as they go away, as they say, they see it burn. But this is nothing but a fable.
>
> (Theophrastus, *Enquiry into Plants* 9.5.2[23])

Such stories were important to report because they apparently circulated widely. Both the historian Herodotus and the philosopher Aristotle report their own version of the story. According to Herodotus, whose source was the Phoenicians who traded the plant, *kinnamōmon* was protected by great birds nesting in *kinnamōmon* nests, located on inaccessibly high cliffs. In order to gather the twigs, the Arabians tempted the birds with pieces of meat (*Histories* 3.111). Their nests could not bear the weight of such offerings and fell to the ground, wherefrom the Arabians collected the 'cinnamon.' According to Aristotle, on the other hand, the nests were located in high trees, wherefrom the Arabians shot the cinnamon birds with arrows laden with lead (*Enquiry into Animals* 8.13, 616a7–13).[24]

Why such differences between the stories of Herodotus, Aristotle and Theophrastus? Few in antiquity could claim to have seen Arabian or Indian plants in their native environment, and buyers had to rely on the accounts of travellers (who often did not really know what they were looking for) or sometimes unscrupulous merchants. In fact, Pliny blamed merchants for inventing such stories to increase the price of their wares (*HN* 12.85). However, blame should not only be laid at the merchants' door. The differences between the three accounts are a good indication of the vicissitudes of oral transmission: stories change as they are told over and over again, and mythologising tendencies are common in oral transmission. That is, the more a story is told, the more 'motifs', which can be transferred from one tale to another, become apparent.

We end this section, with another source of oral information: gods. The compiler of the *Cyranides* mentions in his prologue (paragraph 4) that his knowledge of plants, stones and animals comes from the God, that is, Hermes. Thessalus/

Harpocration, in the preface to his *On the Virtues of Plants*, a letter to 'Caesar Augustus', tells the intricate story of his path to learning. He had travelled from his home in Asia to Alexandria in Egypt to study with the most notorious teachers and physicians (paragraphs 3–5). Before returning home, he went to browse the Alexandrian libraries for medical material, and found the books of the legendary king Nechepso 'dealing with twenty-four medical treatments of the whole body and of every condition according to the signs of the zodiac through both stones and plants'. He boasted about his discoveries, but miserably failed to cure anyone (paragraphs 6–8). In a state of great despondency, and on the verge of suicide, he started wandering around Egypt. Reaching the city of Diospolis (see Map 2), he befriended high priests (paragraphs 9–14). One of them accepted to facilitate a visitation (*episkepsis*) by the god Asclepius (paragraphs 15–23). The god addressed him as follows:

> King Nechepso, a man who is most sound of mind, and honoured with all virtues, was not fortunate enough to obtain what you seek to learn from an utterance of the god. Endowed with a good natural ability, he observed the sympathies of stones and plants [with the stars], but he did not know the right times and places to gather plants. For all seasonal fruits grow and wither under the influence of the stars.
>
> (Thessalus/Harpocration, *Letter to Caesar Augustus* 27–28[25])

The god's revelation is far superior to the writings of one of the most famous astrologers of antiquity: the legendary king Nechepso. Ambivalence towards writing is not, however, limited to the astrobotanical tradition. It is relatively common in ancient technical and philosophical writing in general, and in texts concerned with plants in particular, as we will now discuss.

2.4 Written sources

Some ancient botanical authors refer to written evidence on a regular basis, while others do so only on occasion. In the former category, we can place Pliny the Elder, Columella, Varro and Nicolaus of Damascus; in the latter, Palladius, Dioscorides and, in general, the compilers of herbals. Theophrastus, who was writing at a much earlier time, had far fewer written sources to rely on, but nevertheless refers to them relatively frequently – but not consistently. He referred to various '*phusiologoi*', that is, philosophers (see p. 56 for references); to the poets Homer (see e.g. *HP* 9.15.1), Hesiod (see e.g. *HP* 7.13.3), Musaeus (see *HP* 9.19.2), Cheremon (*HP* 5.9.4), and Aeschylus (*HP* 9.15.1); to two authors of what appears to have been agricultural manuals: Androtion (*HP* 2.7.2) and Chartodras (*HP* 2.7.4); and to the historian Androsthenes (*CP* 2.5.5). On the other hand, he never mentioned Aristotle, with whom he worked closely or to his contemporary and co-citizen of Eresus, Phanias (see Keyser 2008d for references). This other peripatetic philosopher wrote a treatise *On Plants* to which

Athenaeus (second-third century CE, see p. 58) refers on several occasions (Phanias' fragments are collected in Wehrli 1969). It was customary in the classical world not to refer by name to people too close to one's own. Diogenes Laertius, in his *Life of Theophrastus* (5.37), however, quotes from a letter allegedly written by Theophrastus to Phanias, possibly hinting at collaboration between the two philosophers. Writing in the same Peripatetic tradition as Theophrastus, but several centuries later, Nicolaus of Damascus mentioned numerous sources, mostly philosophical in nature: Plato, Aristotle, Theophrastus and several pre-Socratic philosophers.

Dioscorides, on the other hand, rarely named authorities in the body of his text, and the number of those he quoted is rather limited: Hippocrates (see e.g. *MM* 3.59.1), Crateuas (see e.g. *MM* 2.127); Andreas (see e.g. *MM* 4.118, Andreas of Carystus was a third-century BCE physician, see Irby-Massie 2008a for references); Nicander of Colophon (see e.g. *MM* 3.29.1); and Theophrastus (see e.g. *MM* 3.74.4). This is not surprising when one considers how critical he was of his predecessors in the preface (see Chapter 1). There, he condemned the 'ancients' (Iollas of Bithynia, Heracleides of Tarentum, Crateuas the root-cutter, and Andreas the physician) for being incomplete, and the 'moderns' for being inaccurate and for writing theory-laden descriptions of plants. His 'moderns' are Julius Bassus, Niceratus, Petronius [Musa], [Sextius] Niger, and Diodotus (see Scarborough 2008f, 2008h, 2008i, 2008j, 2008l for references to these authors), whom he all classifies as 'followers of Asclepiades', that is, Asclepiades of Bithynia, a famous physician active at Rome, who argued that the cosmos was made of corpuscles (or particles, *onkoi*) (see Scarborough 2008d for references). Of these particles, Dioscorides wrote:

> They did not measure the properties of drugs through experience (*peirai*), but talking vainly about causes, they attributed to each of them differences in particles (*onkōn*), confusing one for the other above everything.
>
> (Dioscorides, *Materia Medica*, preface 2)

Dioscorides' reasons for not naming his predecessors in the body of his *Materia Medica* are therefore clear: his account is based on his vast experience, unlike those of his predecessors, whose knowledge is bookish, incomplete and inaccurate. Even Crateuas he accused on one occasion of relying on hearsay (*historias erein*) when reporting that the male species of a plant helps produce boys, and the female one girls (*MM* 3.125). We should, however, be careful not to take Dioscorides' claims too seriously. While Dioscorides singles out Sextius Niger as one of the worst botanical authors, even providing examples of his mistakes (preface 3), he certainly used his works as a source for the *Materia Medica.* Indeed, as first noted by Max Wellmann (1889), there are numerous examples of parallel passages in the works of Dioscorides and Pliny; while the former is silent as to his sources, the latter acknowledges Sextius Niger (see e.g. *HN* 32.26 = *MM* 2.24). Wellmann even went as far as to claim that 'Dioscorides is nothing other than a compiler of the first century after Christ' (1889: 548); although he later changed his opinion of the ancient pharmacologist while editing his text.

We note that not listing sources is a common characteristic of herbals such as Dioscorides' *Materia Medica*. Thus, Galen who was not usually averse to name-dropping, and who mentioned several sources in the preface to *Simples*, named very few in the actual chapters on plants, which are ordered in the same way as a herbal. Other Greek herbals, for instance the herbal sections in the medical works of Oribasius, Aetius and Paul of Aegina contain very few references to sources apart from Galen and Dioscorides. The same can be said of Latin herbals: pseudo-Dioscorides' *Medicines made from Female Herbs*, pseudo-Asclepius' *Herbarius* and pseudo-Galen's *Alphabet* mention no sources whatsoever.

Palladius, who was writing in the agronomical tradition rather than the herbal one, names very few written sources, and does so on rare occasions: the authors he names are Columella (see e.g. *OA* 11.8.2), Gargilius Martialis (see e.g. *OA* 2.15.6) and 'the Greeks', whom he lumps together (see e.g. *OA* 3.24.1). As mentioned earlier, Palladius emphasised his practical experience of agriculture in general, and of plants in particular. He had, however, one further reason not to quote his predecessors too often: their style was at times far too elaborate. In his preface, he writes:

> Common sense means that you must first assess the type of person you intend to teach. Indeed, if you are instructing a farmer, you should not emulate the skills and eloquence of a rhetorician, as most [instructors] have done in the past. By speaking eloquently to country people, they have managed to make their teaching incomprehensible, even to the most educated. But we must avoid procrastinating in our preface, so as to not imitate those we criticise.
>
> (Palladius, *The Work of Agriculture*, preface 1)

Palladius wanted to write a text which was truly useful to farmers (literate ones, that is), not a rhetorical work. His criticism was most probably directed at Columella, whose style can at times be rather wordy and convoluted, and who refers to rhetoricians in his own preface (*RR* 1 preface 3). As we saw earlier, Columella had extensive farming experience, and made this fact clear in his works. He did, however, also encourage his readers to study the works of the ancients, with the following proviso:

> The works of these authors teach rather than make the craftsman. It is practice (*usus*) and experience (*experientia*) that rule in the arts, and there isn't a discipline in which one is not taught by his own mistakes.... And nobody will be a master of agriculture by reading these theories, unless he is willing to practice them and he has the resources to do so.
>
> (Columella, *On Agriculture* 1.1.16)

In book one, Columella listed forty-three Greek authorities who wrote on husbandry, some famous others 'of lesser fame' (*RR* 1.1.10); ten Latin ones, and the Carthaginian Mago. He noted that there were many other Carthaginian writers on husbandry, but that 'many of them are assailed as erroneous by our farmer' (*RR* 1.1.6). Beyond the preface, Columella mentioned mostly Roman authorities, often to disagree with them: the first-century BCE politician and agronomist Tremellius Scrofa (see e.g. *RR* 2.1.2; see Thibodeau 2008f for references); Marcus Terentius [Varro] (see e.g. *RR* 3.3.2); the first-century CE encyclopaedist Cornelius Celsus (see e.g. *RR* 2.2.24; see Stok 2008a for references); Cato (see e.g. *RR* 4.11.1); the second-first century CE agronomists Hostilius Sasernas, a father and his son (see e.g. *RR* 2.13.1; see Thibodeau 2008c for references); and the polymath Julius Hyginus (see e.g. *RR* 3.11.8; see Cusset and Thibodeau 2008 for references). Vergil he admired without reserve, calling him the 'truest of poets' (*RR* 1.4.4). Among the Greek authorities, Columella mentioned Theophrastus, but indicated that his works were not particularly well suited for farmers (see *RR* 1.5.1).

It appears, then, that Columella was more familiar with some of the authors he listed in his preface than with others. Some he clearly knew second-hand or in a limited way. For instance, he lists King Attalus and King Philometor as two separate people, when they were actually the same king: Attalus III Philometor of Pergamum (Pliny makes the same mistake in the first book of the *HN*). It is possible that Columella borrowed this list of authorities. Indeed, a very similar list had appeared in Varro's preface. There Varro had indicated that there had been more than fifty Greek writers on the topic of agriculture, either in prose or in verse, including several kings. These had been surpassed by Mago of Carthage, whose twenty-eight books in Punic were translated into Greek by Cassius Dionysius of Utica (Varro, *RR* 1.1.7–10; see Thibodeau 2008b for references).

Pliny too listed his sources in book one of the *Natural History*, which serves as a Table of Contents to the entire work. He divided them into Roman and foreign sources. In the body of the *Natural History*, he gave a short history of the art of medical botany in book twenty-five, and dropped names constantly in all books dealing with plants. Theophrastus, 'one of the most celebrated Greek authors' (*HN* 15.1), was one of Pliny's most important botanical sources. The encyclopaedist took great pains to be familiar with Theophrastus' works, sometimes even consulting various copies of the *Enquiry into Plants*. Indeed, Pliny wrote that Theophrastus did not specify the location of a plant 'at all events in the copies of his work that have come into my hands' (*HN* 19.32). However, in the pharmacological books, as noted by Robert Sharples, Pliny's knowledge of Theophrastus appears to be second-hand (1995: 127). There Pliny seems to have relied heavily on former compilations. The other botanical sources of Pliny are extremely varied, comprising philosophers, poets, agronomists, geographers/historians, medical writers, kings, the Magi, and even epigraphical sources (an inscription in verse preserved in the temple of Asclepius in Cos is given at *HN*

20.264). Interestingly, these various sources are often mentioned on the same footing: in Pliny's eyes a poet could offer as valid a piece of information on plants as Theophrastus. For instance, at the beginning of book fifteen, on the olive tree, Pliny mentions Theophrastus, Fenestella (a Roman historian active under Tiberius), Hesiod, Fabianus (Roman philosopher, first half of the first century CE), and Vergil in the same paragraph on the habitat and cultivation of the olive:

> Theophrastus, one of the most famous Greek authors, active around the year 340 of Rome [314 BCE], affirmed that the olive only grows within forty paces of the sea, while Fenestella tells that during the reign of Tarquinius Priscus, in the year 173 of Rome [581 BCE], it did not grow at all in Italy, Spain or Africa. Now, however, it has reached even beyond the Alps and well into Gaul and Spain ... Also Hesiod, who considered that teaching agriculture was among the most important things in life, argued that nobody ever gathered fruit from an olive tree sown during his own life – so slow the process was at the time; whereas now olive trees bear even in nurseries, and after they have been transplanted, olives are gathered from them the next year. Fabianus argues that the olive cannot grow in very cold or very hot places. Vergil said that there are three types of olive: the *orchis* (testicle-shaped), the *radius* (distaff-shaped), and the *posia*.
>
> (Pliny, *Natural History* 15.1–4)

Pliny's sources in this passage are diverse indeed: some sources were focused on plants (Theophrastus and to a certain extant Virgil), others were probably not (the historical writings); some were written in verse, others in prose; some are now lost, others are preserved. In the next section, we are now going to present some of these written sources that we did not introduce in Chapter 1, either because they are not primarily focused on plants or because they are lost. We shall consider the following categories: historical and geographical works; medical works with a focus on plants (herbals); philosophical works; and poetic works. We understand that these categories are artificial. We also make no claims to be exhaustive. Rather, we attempt to show the variety of interests in plants in the ancient world.

2.4.1 Historical and geographical works

We start this overview of the written sources used by botanical writers by looking at 'historians' for one particular reason. Both Theophrastus, the 'father of botany', and Herodotus, the 'father of history', wrote *Enquiries*, *Historiai*. While the modern English word 'history' refers to the study of the past, the Greek word from which it derives, *historia*, had a much broader meaning: it encompassed all systematic enquiries. In many ways, the methods of Herodotus and Theophrastus are similar. We have seen that they had access to similar stories, but there is much more. Both practised a form of ethnography, making enquiries of native people, comparing various accounts and pointing out inconsistencies.[26] Thus, in his section on Cyrene, Herodotus reports different accounts of the establishment of this colony (the Theran and the Cyrenaean versions: *Histories* 4.150–159); and Theophrastus, in his report on the famous Cyrenaic plant silphium, compares two descriptions of the plant (the Cyreanean and that of unnamed 'others') and flags up various inconsistencies, concluding that 'which one of these versions holds true remains to be examined' (*HP* 6.3.2–6).

Of course there are also some differences between Herodotus' and Theophrastus' methods. Herodotus is more interested in the marvellous aspects of the plants than Theophrastus. Herodotus spends more time telling plant stories, such as the one about the *kinnamōmon*-bird – stories which no doubt would have been highly entertaining to his audience. However, even though plant stories had the potential to amaze people, or to inform them as to the strange habits of foreigners, one should not trivialise their role in historical narrative. Plants were extremely important from an economic point of view in the ancient world, and could even be listed as a reason for attempting the conquest of a land. For instance, the historian Arrian wrote that Alexander contemplated conquering Arabia (although he never had a chance to do so) for the following reason:

> Then the prosperity of the country [Arabia] incited him, since he heard that in their oases cassia grew, and from the trees came myrrh and frankincense; and from the bushes, cinnamon was cut; and that from their meadows nard grew self-sown. Then there was also the size of their territory, since the sea-coast of Arabia was reported to him to be not less long than that of India, and that there were several islands adjacent.
>
> (Arrian, *Anabasis* 7.20)

In this passage, the botanical incentive is mentioned before the size of the territory; and the inhabitants of the land are not even cited. Conquering Arabia would have allowed Alexander to control the lucrative trade in 'Arabian' spices, spices that were essential in religious practice.[27] Further, plants were symbolically very important: appropriating plants equated with conquering the country down to its roots. In Persia and Babylonia, large gardens or parks were one of the symbols of kingship. By seizing plants, Alexander and other kings signified their full domination over a land. One should be careful, therefore, not to see Alexander and the people who accompanied him as 'scientists', with systematic interests in cataloguing plants and animals (see Chapter 6 for more detail on plants and conquest). In times of peace, knowledge of foreign plants was also useful to prevent dishonest merchants from tricking gullible buyers into purchasing counterfeited and falsely labelled goods.

Because they were so significant from an economic and symbolic point of view, and because plant stories had the potential to entertain and amaze, historians and geographers throughout antiquity mentioned them in their accounts. For instance, Xenophon describes various plants encountered by the army of Ten Thousands in their march through Asia (see e.g. *Anabasis* 1.2.22, Amigues 1995a); and the Greek-writing traveller Pausanias (second century CE), in his *Description of Greece*, often mentions the trees he saw in various sanctuaries and other sites (see e.g. 2.1.7). Among the lost authors, it is worth mentioning King Juba of Mauretania (turn of the first century BCE–first century CE, see Zucker 2008a for references) who discussed plants on numerous occasions in his historical/geographical works on Libya and Arabia, and who is an important source for Pliny the Elder. Juba also wrote a short treatise, still extant in Pliny's day, on the plant *euphorbia* (spurge), named after the famous physician Euphorbus (*HN* 25.77; see Totelin 2012a on the discovery of *euphorbia*).

2.4.2 Medical works and herbals

Euphorbus was the brother of the physician to the emperor Augustus, Antonius Musa. As mentioned in Chapter 1, the name of Musa is attached to a short treatise on the herb betony, which details the healing powers of the plant in the 'head-to-toe' order, and which is generally transmitted alongside pseudo-Apuleius' *Herbarius*. That attribution is most certainly spurious, but the treatise can be seen as a representative of a common genre in the ancient world: the short medical treatise devoted to a single plant. For instance, a certain physician Chrysippus wrote a book on the cabbage (*HN* 20.78), and the Methodist physician Themison (first century BCE, see p. 22 of this book) one on plantain (*HN* 25.80), of which only short fragments survive. These treatises included short botanical descriptions of the plant in addition to medical indications. Most of these short treatises, however, have fallen victim of the vicissitudes of manuscript transmission and are no longer extant.

These short texts existed alongside longer medical herbals, more exhaustive in scope. Several of these herbals have been preserved and are an important source for the history of ancient botany (see Chapter 1). Many others are now lost. For instance, Diocles of Carystus (fourth century BCE, see Manetti 2008 for references) wrote a *Rhizotomikon* (a 'root-cutting' treatise), which appears to have contained short descriptions of plants, their natural habitats and medical applications: 'Diocles, in the *Rhizotomikon* says that *erinon* is similar to *ōkimon.* It helps against [bites] of wild beasts. It grows near rivers, springs and sunny spots' (Diocles fr. 204 van der Eijk = Scholia to Nicander's *Theriaka* 647, pp. 241–242 Crugnola[28]). Wellmann (1898a: 23–31) suggested that Diocles had been an important source for Theophrastus, Dioscorides and Pliny the Elder, a claim that is difficult to prove, especially since neither Theophrastus nor Dioscorides name him.

Crateuas the root-cutter also wrote a *Rhizotomikon* which, as already mentioned, Dioscorides used as a source. According to Pliny, the work was illustrated in colour with illustrations drawn from the objects (*HN* 25.8, see Chapter

4 for more detail). Crateuas' correspondent, King Mithradates, apparently also wrote medical treatises. According to Pliny, he collected detailed knowledge of herbs from all his subjects, and recorded their medical properties in books, of which there was a bookcase full. The Roman general Pompey, after he had vanquished Mithradates, seized the king's medical treatises and had them translated into Latin by his freedman Lenaeus – 'this great victory therefore was as beneficent to life as it was to the State' (*HN* 25.5–7). Less known are the circumstances in which were written works such as *On Vegetables* by Euthydemus the Athenian (quoted in Athenaeus, *Deipnosophists* 2.58f); *Hyacinth* by Andreas; and *On Vegetables* by Epainetes (both quoted in Scholia to Nicander's *Theriaka* 585a).

No medical authority, however, is given more importance in ancient botanical texts than Hippocrates. Throughout antiquity and beyond, the physician's herbal lore was held in high esteem. He is quoted or mentioned as an authority on plants by Galen, Dioscorides, Pliny, Varro, Columella, Palladius and Pliny. The reader of the Hippocratic Corpus might find this emphasis on Hippocrates' herbal lore slightly puzzling. While, as mentioned in Chapter 1, Hippocratic writers mention numerous plants in their pharmacological recipes and dietetic prescriptions, they give almost no descriptions, and very few indications as to where plants grow. No preserved Hippocratic text is strictly herbal in character. While it is possible that some Hippocratic material may be lost, it is safer to assume that the criteria for assuming the status of botanical authority were different in the ancient world as they are today. One did not need to write a herbal to be considered an authority on botany; one had to be able to use plants to good effect or have a good understanding of their physiology, as did the ancient philosophers, to whom we now turn.

2.4.3 The works of philosophers, mystics and magicians

The earliest Greek philosophers, the so-called pre-Socratics, clearly had an interest in the origin of plants, the place of plants in the order of things, the similarities (or lack thereof) between plants and animals, the generation of plants, and other difficult questions. Theophrastus discussed the opinions of the following pre-Socratic philosophers: fifth-century Anaxagoras of Clazomenae (see e.g. *CP* 1.5.2, see Graham 2008a for references); fifth-century Empedocles of Acragas (see e.g. *CP* 1.22.2, see Trépanier 2008 for references); mid-fifth century Pythagorean Menestor of Sybaris (see e.g. *CP* 1.21.6, see Zhmud 2008b for references); mid-fifth century Pythagorean Hippon of Croton (see e.g. *HP* 1.3.5, see Zhmud 2008a for references); mid-fifth century Leophanes (see *CP* 2.4.12, see Marsilio 2008b for references); late fifth-century Diogenes of Apollonia (see e.g. *HP* 3.1.4, see Graham 2008b for references); fifth–fourth century Democritus of Abdera (see e.g. *CP* 6.1.6, see Englert 2008 for references); fourth-century Cleidemus of Athens (see e.g. *HP* 3.1.4, see Marsilio 2008a for references). Nicolaus of Damascus' *On Plants* too is a rich source for the opinions of the pre-Socratic philosophers.

Unfortunately, the original works of the pre-Socratics are now lost, and it is impossible to know whether Theophrastus and Nicolaus at times misrepresented the philosophical views of their predecessors. One should not forget that Theophrastus was a philosopher working in the same argumentative tradition as the pre-Socratics, Plato (who discusses plants in the *Timaeus*) and Aristotle. The writings and opinions of the pre-Socratics could lend themselves to a variety of interpretations and re-appropriations, some of which led to the creation of entirely new traditions. Thus, magicians, astrologers and Hermetic writers appropriated for themselves the theories – and names – of Democritus, Empedocles and Pythagoras. By the time of Pliny, writings on magical plants circulated under the names of Pythagoras and Democritus: 'These [magical plants, *magicis*] were first brought to the notice of our part of the world by Pythagoras and Democritus, who followed as their authority the Magi' (Pliny, *Natural History* 24.156).

> After Homer, Pythagoras, famous for his wisdom, was the first to write a book about the powers of plants, attributing their discovery and origin to Apollo, Aesculapius, and in general to the immortal gods. Democritus composed a similar work. Both travelled to the Persian Magi, and those of Arabia, Ethiopia and Egypt, and the ancients were so astounded by these works that they maintained even unbelievable facts.
>
> (Pliny, *Natural History* 25.13)

Pliny himself was aware of the authorship issues surrounding some of these magical books: he noted that the works of Pythagoras on plants were sometimes ascribed to the Hellenistic physician Cleemporus (*HN* 24.159, see Irby-Massie 2008b for references). While Pythagoreans certainly wrote on the topic of botany, Pythagoras himself wrote nothing, even though numerous writings were attributed to him in later antiquity. As for Democritus, his name sometimes became associated with that of the mysterious Egyptian Bolus of Mendes, probably active in the second century BCE (see Hallum 2008 for references). Palladius wrote in a paragraph on veterinary medicine of 'the notable Egyptian author [Bolus], a citizen of Mendes, whose handbook is wrongly published under the name of Democritus' (*OA* 14.32.6). According to the tenth-century encyclopaedia, the Souda, Bolus was a Democritean who composed, among other things, a book on *Naturally Potent [Drugs?]*, and one *On Sympathies and Antipathies*, both of which are likely to have included information on plants (Souda, s.v. *Bōlos*, B481 Adler).

That authors of magico-botanical works chose the names of famous ancient philosophers as their pseudonyms is telling: they wanted to place themselves in that tradition. And, as noted by classicist Guy Ducourthial (2003: 99), there are indeed similarities in their approach to the cosmos, in particular in their positing of antipathies and sympathies between the constitutive elements of the cosmos. Most of this magico-philosophico-botanical tradition is now lost to us, and only known through fragments in the works of authors such as Pliny, Columella and Palladius who quoted them on various occasions.

As for philosophers in a more restricted understanding of the word, throughout antiquity they kept debating over plant matters. We shall mention various philosophers in this book, and we are content here with giving three examples. The Neo-Platonist Porphyry (*c.*232/234–*c.*305 CE, see Karamanolis 2008 for references) in *To Gaurus* (*Ad Gaurum*), a treatise on embryology, makes numerous mentions of plants, seeds and of grafting. Plutarch, the historian and philosopher, in his *Table Talk* discussed several botanical topics: for instance 'whether ivy is of a hot or cold nature' (3.2, 648b–e); 'why truffles are thought to be produced by thunder' (4.2, 664b–666d); or 'why women do not eat the middle part of lettuce' (4.10, 672) (unfortunately, the answer to that last question is lost). And the sophists in Athenaeus' *Sophists at Dinner* (an account in fifteen books of the discussion held at a fictional banquet, second or third century CE, see the studies in Braund and Wilkins 2000) discussed the classification and properties of fruits and vegetables, quoting various authorities, both lost (as with Phanias) and preserved (as with Theophrastus and Nicander).[29]

2.4.4 Poems

In Chapter 1 we introduced some poetic writings that had a botanical or agricultural focus. However, defining what constitutes a 'botanical focus' is not always an easy task. While poems such as Columella's book ten on horticulture or Palladius' *On Grafting* can certainly be put in that category, what about the *Theriaka* of Nicander, which mention numerous plants but are not centred on them? And should we consider the third-century BCE Theocritus a botanist, as classicist Alice Lindsell did, since he described numerous herbs and trees (eighty-seven in total) in his pastoral *Idylls*, and demonstrated a very good knowledge of flora (Lindsell 1937, see also Fée 1832; Amigues 1996a). Some of this knowledge he may have acquired indirectly from Theophrastus. According to a scholium to *Idyll* 11 (argument d, line 3, p. 240 Wendel), Theocritus studied together with the physician Nicias of Miletus (the dedicatee of the *Idylls*, see Meliadò 2008 for references) under the famous physician Erasistratus of Ceos (see Scarborough 2008g for references), himself a student of Theophrastus.

In addition to asking whether a poem has a botanical focus or not, one could enquire whether a poem attempts to teach botany (whether it is a didactic poem) or whether it merely refers to plants to entertain audiences, to set the scene. Thus, some would place Columella's poem on gardening, Palladius' poem on grafting, and Nicander's works in the 'didactic' category; while they would place Theocritus in the 'entertainment' one. However, we would argue that the ancients 'learnt' about plants from any poem that mentioned them, not merely from those that aimed to teach plant knowledge. In this respect one should note that Homer, who mostly mentioned plants incidentally in the *Iliad* and *Odyssey*, was considered a botanical authority throughout antiquity. His mention of the plant moly, in particular, attracted much scholarly discussion.[30] In a famous scene of the *Odyssey*, Hermes offers Odysseus the plant moly to ward off Circe's baneful drug that had turned his companions into pigs:

> So saying, Argeiphontes [Hermes] gave me the remedy,
> Drawing it from the earth, and showed me its nature.
> It was black at the root, but like milk was its flower.
> 'Moly', thus the gods call it; it is difficult to uproot
> For mortal men. But gods can do everything.
> (Homer, *Odyssey* 10.390–394)

Homer's moly may be nothing more than a mythical plant, but since the name 'moly' was used in Greek to designate actual plants, much debate ensued. Theophrastus wrote that:

> Moly grows around Pheneus and on the Cyllene [mountain of Arcadia]. They say that it is similar to the plant that Homer described, as it has a round root resembling that of onion, and a leaf similar to that of squill. It is used as antidote and in magical practices. However, it is not difficult to dig up, as Homer has said.
> (Theophrastus, *Enquiry into Plants* 9.15.7)

Theophrastus, however, might have been too literal in interpreting the Homeric verses. The poet was merely referring to the difficulty accompanying the gathering of all magical plants – that difficulty disappears when one honours all the necessary rituals. Pliny for his part questioned Homer's description of the plant's flowers, relying on the authority of a herbalist:

> Greek authorities have depicted it with yellow blossom, although Homer described it with white flowers. I have found a physician experienced in herbs who told me that the plant also grew in Italy, and that one could be brought to me from Campania in a few days.
> (Pliny, *Natural History* 25.27)

Like Theophrastus, Pliny was being over-literal in his interpretation of poetic material.

Thus plants of all sorts provided a rich source of inspiration to poets throughout antiquity. Many plants were associated with divinities (olive tree with Athena; pomegranate with Persephone) or had been mythical characters in their previous incarnation (see e.g. Dierbach 1833; Murr 1890; Baumann 2007; Giesecke 2014). For instance, Apollo had turned Daphne into a laurel, a story told for instance by Ovid (43 BCE–17/18 CE) in the *Metamorphoses* (1.452–567). Plants' capacity for renewal and their rootedness was also the topic of numerous poetic metaphors and analogies. Comic poets exploited the sexual connotations of many plants in numerous puns and double-entendres (see Henderson 1991). The comedian Euboulus (fourth century BCE) wrote a play on 'wreath-makers', which no doubt was filled with sexual innuendos. Love poets, starting with Sappho (seventh century BCE), sang in the language of flowers (see e.g. Calame 2006). Later in antiquity, compilers would collect *Anthologies*, literally '*Collections of Flowers*'. The preface to Meleager of Gadara's anthology, titled *Stephanos* (Wreath, *Greek Anthology* 4.1), draws complex analogies between several poets and flowers (Sappho for instance is the rose at verse 6).

2.5 Conclusions

In this chapter, we distinguished three types of sources of plant knowledge in the ancient world: personal expertise, oral sources and written sources. However, it soon became clear that one has to be wary of drawing sharp boundaries between these categories. Some oral information can be written down and become subject to the vicissitudes of manuscript transmission; while written tales can be retold orally, and transformed in the process. It is also important not to make assumptions regarding the origin of ancient botanical knowledge. Knowledge that appears to us 'superstitious' or 'folkloric' does not necessarily originate from illiterate country people.

One of the main themes of this chapter was that of authority: what did it take in antiquity to be seen as an authority on plants? We saw that botanical authors used as their sources people of various levels of education and literacy; people whose interest in plants varied from the highly practical to the highly metaphorical; and people of diverse geographical origin. Plant knowledge, sometimes extremely technical in nature, was widespread in the ancient world. In this context, it is very difficult to draw a boundary between the strictly botanical and non-botanical in our ancient material. In the next chapter, we will discuss how plant classifications often had their origin among people who worked with plants for their livelihood.

Another theme that became apparent in this chapter and that will reappear throughout this book is the link between political and commercial expansion and increasing botanical knowledge. Indeed conquest, colonisation and expeditions to foreign lands were significant means to acquire botanical knowledge in the ancient world. A final theme of this chapter has been wonder. The powers and appearance of plants never ceased to amaze the ancients and inspired some of their most beautiful literary works.

Notes

1 See *HN* 14.1–3, 23.112 and 25.1 for similar claims.

2 See also *HN* 25.38 (Mercury), 25.42 (Achilles), 25.47 (Melampus), 25.10 (Medea), 25.12 (Helen), 25.3 (Chiron). Pseudo-Apuleius' *Herbarius* also notes when a god was involved in discovering a plant, see e.g. chapter 12, p. 45 Howald and Sigerist (Diana).

3 A dichotomy between 'ancients' and 'moderns' is also to be found in Dioscorides, *Materia Medica*, preface 2 and in the pharmacological writings of Galen.

4 On Pliny's notion of nature see Beagon 1992.
5 On the links between empire, conquest and botany in the ancient world, see e.g. Dalby 2000; Macaulay-Lewis 2008; Pollard 2009; Totelin 2012a.
6 See also Varro, *RR* 1.18.7, where the agronomist explains the role of expertise and imitation in farming (*experientiam et imitationem*).
7 There is an extensive literature on the topic of orality and literacy in the ancient world. The two most useful general works remain Harris 1991 and Thomas 1992.
8 Theophrastus also uses the word '*empeiroi*' at *HP* 2.6.1 and 6.3.7. On Theophrastus' *empeiria*, see Strömberg 1937: 22; Wöhrle 1985: 49–51. On *empeiria* in general see von Staden 1975. On Theophrastus' method of research more generally, see Amigues 1999a.
9 See e.g. Greene 1909: 56; Capelle 1954a; Pease 1952; Thomson 1955: 15; Amigues 1999a; Ducourthial 2003: 56.
10 On Theophrastus's travels, see Einarson and Link 1976: xxiii; Amigues 1988a: ix–xvi; Thanos 2005a. On Cyrene in particular, see Capelle 1954b; Amigues 2004a. On silphium, see p. 120 in this book.
11 On scientific enquiry and Alexander's expeditions, see e.g. Bretzl 1903; Joret 1904; Romm 1989; Bodson 1991; Amigues 1996b; Karttunen 1997: chapter 5.
12 On the persea, see Amigues 1986.
13 On Palladius' personal experience, see Fitch 2013: 11.
14 The Greek word used here is *dendron*, whose usual translation is 'tree'. However, in some cases, *dendron* had a broader meaning, and can be translated as 'plant'.
15 As pointed out above, the extent of their travels is matter of debate, and here we focus on peoples they probably met, rather than, for instance, the 'Indians' to whom both Theophrastus and Dioscorides refer on various occasions but probably never met.
16 On this phrase see Nutton 1991: 25; Totelin 2012b: 309.
17 Gardeners: Greek: *kēpouroi*, see e.g. Theophrastus, *HP* 7.2.5; Latin: *holitores*, see e.g. Columella, *RR* 11.3.3. Farmers: Greek: *geōrgoi*, see e.g. Theophrastus, *HP* 2.6.2; Latin: *agricolae*, see e.g. Palladius, *OA* 3.17.5. Vine-layers: Latin: *mergi*, see e.g. Palladius, *OA* 3.16.2. Pruners: Latin: *putatores*, see e.g. Palladius, *OA* 1.35.1. Beekeepers: Greek: *melittourgoi*, see e.g. Theophrastus, *HP* 6.2.3; Latin: *apiarii*, see e.g. Pliny, *HN* 21.56. Shepherds or herdsmen: Greek: *nomeis*, see e.g. Theophrastus, *HP* 9.17.1 or *poimenes*, see e.g. Dioscorides, *MM* 3.90. Wreath-makers: Greek: *stephanoplokoi*, see e.g. Dioscorides, *MM* 3.75; Latin: *coronarii*, see e.g. Pliny, *HN* 21.68. Woodcutters: Greek: *hulotomoi*, see e.g. Theophrastus, *CP* 1.5.5. Joiners/carpenters: Greek: *tektones*, see e.g. Theophrastus, *HP* 5.1.12. Torch-makers: Greek: *daidourgoi*, see e.g. Theophrastus, *HP* 3.9.3. Master builders: Greek: *architektones*, see e.g. Theophrastus, *HP* 5.5.4; Latin: *artifices*, see e.g. Pliny, *HN* 13.99. For references to people involved in the drug trade, see p. 43. Merchants: Greek: *emporoi*, see e.g. Dioscorides, *MM* 1.13; Latin: *negotiatores*, see e.g. Pliny, *HN* 18.225. On Theophrastus' *oreitupoi*, a word whose translation is difficult, see Blanc 1996.
18 On these professionals, see e.g. Riddle 1985: 4–6; Korpela 1995; Scarborough 1987; Scarborough 1991: 144; Boudon-Millot 2003; McNamara 2003; Amigues 2006: xxxii–xxxv; Samama 2006; Repici 2006; Totelin forthcoming a. On the ancient medical marketplace more generally, see Nutton 1992.
19 Lived at the court: Massar 2005: 227; helped prepare antidotes: see for instance Debru 1996: 27; was personal physician to Mithridates: Scarborough 2012a. Marasco on the other hand, refuses to see him as a court physician (1996: 457). For the fragments of Crateuas, see volume 3 of Wellmann's edition of Dioscorides (1914: 139–146). Nutton has identified an unknown fragment of Crateuas's book three of the *Rhizotomikon* in *On the Virtue of the Centaury* (2010: 218–219).
20 See Boudon-Millot 2003: 119. Further in the preface, Pamphilus is accused, together with Andreas, of being a charlatan (*Simples* 6, preface, 11.796 Kühn).

21 A translation of the Greek magical papyri is available: Betz 1992. See p. 95 for this invocation.
22 These rituals are studied in detail by Armand Delatte 1961 and Guy Ducourthial 2003. See also Scarborough 2006a. On the gathering of the mandrake, see Patera 1994.
23 Scholars have debated endlessly over the identification of the ancient *kinnamōmon*. De Romanis 1996 avoids the debate, but thinks the word *kinnamōmon* designates a plant from Oriental Africa. See contra Marganne 1996, who notes that Theophrastus believed *kinnamōmon* came from India but was traded by the Arabians, and who argued that ancient *kinnamōmon* is our cinnamon.
24 See also Julius Africanus, *Cesti* F75 (p. 187 Wallraff) where Herodotus is quoted and criticised. On the stories surrounding the *kinnamōmon*, see Detienne 1972: 17–57; Lecocq 2009. On the plant-lore of Herodotus, see Kanngiesser 1910; Forster 1942.
25 Galen also mentions Nechepso, the legendary Egyptian astrologer-king, in relation to the stone jasper (*Simples* 10.2.10, 12.207 Kühn); while Aetius gives a herbal recipe by the king in book one of the *Medical Collection* (1.38, CMG 8.1, p. 40 Olivieri).
26 On Herodotus' historical method, see e.g. Lateiner 1989.
27 We use quotation marks in 'Arabian' because not all these plants grow in Arabia.The literature on these plants and their trade is immense, see in particular Raschke 1978; Miller 1969; Groom 1981; the essays edited by Avanzini 1997; Dalby 2000.
28 See van der Eijk 2001: 390 for a commentary to this text.
29 See Sharples 1995: 127 for Athenaeus' method of quotation of Theophrastus.
30 On the plants in Homer, see Euchholz 1848; Fellner 1897; Forster 1936. The identification of moly has been a topic of scholarly interest for centuries, see e.g. Siber 1699; Wedel 1713; Triller 1716; Dorie 1967; Amigues 1995c.

3

ORGANISING THE VEGETABLE KINGDOM

3.1 Introduction

There is probably no better introduction to the issue of plant classification in the Graeco-Roman world than the following fragment of a play by the comedian Epicrates (fourth century BCE). One character (B) tells his friend (A) about a discussion he heard at the Academy. Plato's students were attempting to determine the place of the pumpkin (*kolokuntē*) in the vegetable kingdom: was it a vegetable, a grass or a tree?[1]

A: What about Plato and Speusippus and Menedemus? What are they debating nowadays? What thought, what idea is discussed in their circle? Tell me wisely, if you come with any knowledge. By the land, do.

B: At the Panathenaea I saw a group of young men ... at the gymnasium of the Academy I heard these unutterable, extraordinary words. For they were defining nature and separating into categories the world (*bion*) of animals, the nature (*phusin*) of trees, and the kinds (*genē*) of vegetables (*lachanōn*). And among other things, they were trying to determine the kind to which the pumpkin belongs.

A: And how did they classify it? To what kind does this plant (*phuton*) belong? Do tell me if you truly know.

B: Well, at first they all sat in silence, and with their heads down they pondered for a long time. Then suddenly, while the young men were still thinking with their heads between their hands, one said that it was a round vegetable (*lachanon*), another said it was a grass (*poia*), and a third a tree (*dendron*). On hearing that, a physician from the land of Sicily eructed/farted at them, as if they were raging mad.

A: They must have got terribly angry at the impertinence and shouted at him? For to do such things in public places is very rude.

B: It did not bother the young men at all. And Plato, who was nearby, very gently, without anger, told them to start from the beginning and define the species to which the pumpkin belongs. So they started again their enquiry.

(Epicrates, fr. 10 Kassel and Austin = Athenaeus, *Deipnosophists* 2.59d–f)

We would argue that Epicrates was well informed and that his joke works on various levels. First, pumpkins were considered funny in the ancient world (see Todd 1943). The Greeks and Romans had various pumpkin jokes, not the least the title of Seneca's work, the *Apocolocyntosis (Pumpkinification) of the Emperor Claudius*. Second, the list of Epicrates' botanical categories seems incongruous: surely the pumpkin could not be a tree;[2] and wasn't it too big to be a herb? Clearly, the category 'vegetable' is the most obvious for the pumpkin, and that is the category Epicrates first mentions. But, in antiquity as today, 'vegetable' was a tricky category. It is based on edibility rather than on any morphological characteristic. And then, when one chooses the pumpkin to exemplify the category of the vegetable one does run into particular troubles, because it is both sweet and savoury.

One may be tempted to react like the Sicilian when dealing with the classificatory efforts of the ancient Greeks. They can appear very abstract and detached from reality. That being said, deciding upon classificatory criteria is no easy task, and many attempts were made at finding useful classificatory principles for plants before the current system became accepted. This system divides plants on the basis of their reproduction, separating them into flowering plants (also known as spermatophytes, that is, seed-producing plants) and non-flowering plants, which include algae, bryophytes and pteridophytes (for an introduction to modern classifications, see Simpson 2010: unit two).

As we will see, the very definition of a 'plant' has changed with time, with significant modifications brought to that definition in the twentieth century. In this chapter, we first examine the place of plants in the *scala naturae*, the 'great chain of being', as well as the actual definition of a plant in antiquity. We then study various systems of classification used by the ancients. We start that examination with a discussion of the notions of *eidos* (species) and *genos* (genus) in ancient botanical texts. We then turn to overarching classificatory systems in the works of Theophrastus (division into four classes according to parts of plants); Dioscorides (classification according to effect on the body); Galen (alphabetical classification); and astrobotanical treatises (classification following the planets or the zodiac). We end this chapter with some considerations of organisms that were counted among 'plants' in the ancient world but are no longer so: fungi and lichens.

3.2 Plants in the order of things

All ancient authors argued that plants were alive. Yet, the question of what constitutes a plant was a difficult one. Today, a plant is defined as an organism that can photosynthesise. Here is the *Oxford English Dictionary* definition:

> a. *gen.* and *Biol.* A living organism other than an animal, able to subsist wholly on inorganic substances, typically fixed to a substrate and moving chiefly by means of growth, and lacking specialised sensory

> and digestive organs; *spec.* (more fully *green plant*) such an organism belonging to a group (the kingdom Plantae) which comprises multicellular forms having cellulose cell walls and capable of photosynthesis by means of chlorophyll, including trees, shrubs, herbs, grasses, and ferns (the vascular or higher plants), and also mosses and liverworts (the bryophytes). Freq. *spec.*: a small (esp. herbaceous) organism of this kind, as distinguished from a tree or shrub; (in informal use) such an organism grown for or known by its foliage or fruit, as distinguished from a 'flower'.

Photosynthesis, however, has only 'recently' been discovered, and the word itself dates to 1893, when American botanist Charles Reid Barnes (1858–1910) coined it (see Gest 2002). As a result of enquiries into photosynthesis, it was found that fungi have non-cellulose cell walls and are not able to produce food by this process. They were, therefore, removed from the plant kingdom by taxonomists and have been placed in a kingdom of their own. Previously fungi were classified, together with the algae, as Thallophyta. Modern research has also shown that the term 'algae' refers to a disparate grouping of several unrelated forms of simple aquatic plants. As a result of these various changes in our understanding of the living world the group Thallophyta has now been discarded (see Ainsworth 1976: 33).

The ancients could not observe photosynthesis. Their definition had to hinge onto something else. That something, for most philosophers, was the question of the soul, and more particularly whether plants were endowed with sensation and motion. This question occupied philosophers from the very first, but since the writings of the so-called pre-Socratic philosophers are fragmentary, we shall start with the earliest fully preserved philosophical texts: those of Plato. The philosopher in the *Timaeus* describes how the Demiurge Gods, after creating the cosmos and animals, created plants:

> They mixed a nature similar to the nature of man with other forms and perceptions so that there would be another type of living being (*zōion*): the plants. These are the trees (*dendra*) and plants (*phuta*) and seeds (*spermata*) which we have taught (sc. improved) through agriculture; before there were only the wild kinds (*genē*), which are older than those of today.
>
> (Plato, *Timaeus* 77a)

Thus for Plato, plants are living beings, with a nature similar to that of man. This, he felt, needed explication:

> For everything that partakes of life (*tēn zēn*) can truly be called a living being (*zōion*). And the living being of which we are now speaking partakes of the third type of soul, the one located between the midriff and the navel, which has no part in opinion (*doxēs*) or reason (*logismou*) or

> mind (*nou*), but only in perceptions of pleasure and pain that accompany them. For this [being] is continuously passive (*paschon*), revolving of itself and around itself, and repelling the motion from without and using its own, and was not provided by nature with the ability to look upon and reflect on its own [affections]. For that reason it is alive and does not differ from a living being (*zōiou*), but is immobile and rooted and fixed, as it is deprived from the power of self-motion.
>
> (Plato, *Timaeus* 77b–c)

Plato is here referring to the lowest part of the soul: the appetitive part, responsible for hunger and sexual desire (the other parts are the spirited part and the thinking part). According to Plato, that part of the soul enables plants to experience pleasure and pain, but only in a passive way. For plants are always passive by virtue of being immobile.

For Aristotle too lack of locomotion was a defining characteristic of plants.[3] Aristotle, however, knew that some animals (shellfish, sponges) were not capable of locomotion: a better definition of motion was needed. Aristotle therefore distinguished four types of motion: 1) change of position; 2) change of state; 3) decay; and 4) growth (*On the Soul* 1.3, 406a13–14). Plants could 'move' in that they grew and decayed. In the following passage, Aristotle clearly classifies growth as motion, and also introduces the three parts of his tripartite soul:

> Indeed it is not the soul in its entirety that is the principle (*archē*) of motion, nor all the parts taken together, but one part may be responsible for growth, that which is found in plants; another part, the sensory, for alteration; and yet another, the intellectual, for movements.
>
> (Aristotle, *Parts of Animals* 1.1, 641b5–8)

In other words, plants have the nutritive part (*threptikon*) of the soul, they can take in food, grow and generate (see Chapter 5 for more details on these functions). Plants, unlike animals, do not have the sentient part (*aisthētikon*) of the soul, responsible for changes of state. Plants are deprived of pain, pleasure, desire, sleep and waking. However, they can be affected by sensibles (see e.g. *On the Soul* 2.12, 424a32–b3). For instance, a plant cannot smell but it can be affected by smells (Panagiotou 1975; Andersen 1976). It follows from Aristotle's definition, that an animal that is deprived of sensation or not yet endowed with sensation (in embryonic state) is 'living the life of a plant' (*GA* 5.1, 778b35), something analogous to sleep, yet not sleep.

Aristotle adds some elements to this definition: plants have as their main function their propagation: 'For in essence, plants have no function or action to perform than the production of seed' (*GA* 1.23, 731a25–26). The *telos* of plants, their final cause, the reason for their existence, is the production of seed. This seed will allow the plant to reproduce, and in many case, will be useful to man. Nature does nothing in vain.

Even though Aristotle achieved a high level of sophistication in his definition of the difference between plants and animals, he knew it was not always easy to distinguish between plant and animal. His solution to this boundary issue was a suspension of judgement: nature progresses continuously, without breaks, from the inanimate to the animate: 'Nature passes continuously from soulless things to animals (*zōia*), passing by living beings (*zōntōn*) which are not animals (*zōiōn*), so that the difference between one class and the next appears extremely small' (Aristotle, *Parts of Animals* 4.5, 681a12–15[4]). Within this great chain of being, Aristotle constantly found similarities between animals and plants, using his observations of physiological processes in animals to explain by analogy physiological processes in plants – and vice versa. The following passage is only one example among many: 'For the roots of plants are equivalent (*analogon*) to the so-called mouth in animals, through which the former take nourishment from the earth, and the latter from each other' (Aristotle, *On Youth and Old Age* 1, 468a10–13[5]). The idea of a continuous *scala naturae* may have its origins in ancient popular thought. Stories of metamorphoses abound in ancient literature, and may have been based on the belief that boundaries between types of living beings were somewhat porous.

All subsequent botanical writers had to respond, in one way or another, to Aristotle's definition of plants. Theophrastus' response was mainly to remain silent. For instance the word 'nutritive' (*threptikon*) only appears once in the works of Theophrastus (*CP* 1.12.5, see Chapter 5 for more detail). Moreover, the philosopher did not discuss the notion of a continuous chain of being, rather stressing the difference between plants and animals, and warning against pushing too far the comparisons of physiological processes in plants and animals. To Theophrastus, the nature of plants was more complex, 'more irregular (*ataktoteran*) and more mixed (*xunkechumenēn*)' than that of other living organisms (*CP* 2.16.6). Theophrastus, then, moved away from debates relating to the soul of plants, to focus instead on classifying plants and understanding their physiology. The philosopher also famously questioned some tenets of Aristotelian teleology, exploring in his *Metaphysics* phaenomena where nature appears to do things in vain. As expressed by historian of philosophy James Lennox, 'Theophrastus' *De causis plantarum*, a book one might expect to be filled with teleological reasoning, is remarkably devoid of it' (1985: 159).[6] As we will see in Chapter 5, Theophrastus' thought on plant generation and fructification is at times teleological, but the philosopher allowed for randomness, arguing in the *Metaphysics* that plants and some lower animals occurred 'spontaneously (*tōi automatōi*) because of the rotation of the whole [i.e. the cosmos]' (*Metaphysics* 10b27–28). In other words, plants were not created for the sake of animals and humans, but appeared randomly.

Nicolaus of Damascus, on the other hand, returned to the Aristotelian view of plant's soul. He opened his treatise with the following statement:

> Life is found in animals and plants, but the life of animals is manifest and clear, whereas the life of plants is hidden and concealed. An inquiry

> and an investigation are needed in order to find the way of truth on these matters.
>
> (Nicolaus, *On Plants* 1.1.1, 815a10–12; translation: Drossaart Lulofs and Poortman 1989: 126)

Nicolaus then expressed his wish to determine whether plants have souls. Like Aristotle, he concluded that plants have the nutritive part of the soul, but not the sentient part, also using the example of shellfish to show how difficult it is to draw the boundary between animals and plants. Nicolaus, however, asserted in no uncertain terms that plants are 'created for the sake of animals only, and animals are not created for the sake of plants' (1.2.53, 817b25–26; translation: Drossaart Lulofs and Poortman 1989: 142). In this respect, he went much further than Aristotle.

Pliny too followed Aristotle in his definition of plants, noting for instance that starfishes and other sea animals have no senses (*nullum sensum*), having the 'same nature as a bush (*frutici*)' (*HN* 9.154). Like Aristotle, Pliny often drew analogies between plants and animals, but did so in a less careful manner, thus producing anthropomorphising accounts of the life of plants.

We do not have the space here to engage in detail with all ancient definitions of plants – a few pointers must suffice. Stoic and Epicurean philosophers, for their part, apparently denied all soul to plants:

> The Stoics and Epicureans [argue that plants] are not ensouled. For some [animals] partake in the appetitive (*hormētikē*) and the desiring (*epithumētikē*) soul; other of the intellectual (*logikē*) soul, but plants are moved fortuitously and not on account of a soul.
>
> (pseudo-Plutarch, *Placita* 5.26, 910b–c)

The Greek-writing Jewish philosopher Philo of Alexandria (turn of the first century BCE and first century CE, see Zucker 2008c for references) achieved a fusion of Judaism and Greek philosophy in his definition of plants. In a manner that recalls Plato, Philo presented his God as a 'Great Planter', whose plant 'does not contain in itself individual species of plants, but rather countless numbers of them shooting up like vine-shoots from one root, that is, the universe' (*Noah's Work as a Planter* 2.2). He then defined plants in a manner that is Aristotelian: 'those [beings] that are governed by a nature without sensation, which are called "plants", have no share of the type of motion that involves change of place' (*Noah's Work as a Planter* 2.13). As for the Neo-Platonic philosopher Plotinus (third century CE, see Karamanolis 2008a for references), he drew a distinction between happiness (*eudaimonia*) and sensation, arguing that 'those who, on account of their lack of sensation (*mē aisthanesthai*), deny [happiness] to plants, run the risk of denying it to all living things (*tois zōois*)' (*Enneads* 1.4.2).

Through the writings of Plato, Aristotle, Theophrastus, Nicolaus, and later philosophers, we get glimpses into the origins of debates on plants among the pre-Socratics. Nicolaus, in particular, gives us useful insights into the issue of the soul of plants:

> Now, Anaxagoras and Empedocles assert that plants have desire and sensation, pain and pleasure, and Anaxagoras <naively> asserts that they are animals and that they feel joy and sadness – and he cites as proof the shedding of their leaves in due season ... Anaxagoras, Empedocles and Democritus maintained that plants possess reason and understanding. But we must refrain from this abominable nonsense.
>
> (Nicolaus, *On Plants* 1.1.3 and 10, 815a17–b19; translation: Drossaart Lulofs and Poortman 1989: 126–128 = DK 31A70)

Pseudo-Plutarch, for his part, associated Plato with Empedocles in their views on the nature of plants:

> Plato and Empedocles argue that both plants and animals are ensouled. This is clear from their shaking and that they stretch their branches, and that they retire when they are under attack (from the woodman) but relax again so much that they draw up weights.
>
> (pseudo-Plutarch, *Placita* 5.26, 910b = DK 31A70)

Unfortunately, these are testimonies from second parties rather than fragments of the original writings of the pre-Socratics. We do not have fragments of Anaxagoras, Empedocles and Democritus asserting that plants had sensation, pain, pleasure, emotions, reason and understanding. Their views may have been much more nuanced than Nicolaus' and pseudo-Plutarch's rather crude statements.

In sum, the questions of what a plant is and how it came to be are complex from a philosophical point of view. No less difficult was the classification of organisms within the 'plant' category. How did the ancients distinguish 'types' of plants? And how did they group types of plants into bigger groups? These are the questions to which we now turn.

3.3 *Diaphorai*, *eidos* and *genos*

Philosophers, starting with Aristotle, classified living beings (animals and plants) into *genē* (plural of *genos*) and *eidē* (plural of *eidos*), terms which one could translate as 'genus' and 'species'. There is, however, no exact correspondence between these ancient categories and the modern notions of genus and species. The reader will find detailed discussions of these issues in numerous modern scholarly works; here we are content with giving a simplified explanation.[7] Aristotle, and Theophrastus after him, usually subordinated the *eidos* to the *genos*. But beyond that, they did not employ the words in any systematic way, as expressed by D. M. Balme:

> There, in [Aristotle's] strictly biological works, although it has been traditional to interpret *genos* and *eidos* as genus and species in many passages, examination shows that there are only a few in which the technical sense if obligatory; in a few more it is possible but not demanded; while in others it even causes confusion and in some it is impossible.
>
> (Balme 1962a: 84)

At times, Theophrastus seems to have used the two words almost as synonyms. Thus, he wrote that 'the beech does not present differences (*diaphoras*), but it is monotypic (*monogenes*)' and a few paragraphs later 'the hornbeam is also monotypic (*monoeidēs*)' (*HP* 3.10.1 and 3.10.3) – *monogenēs* and *monoeidē*s are here synonyms. Dioscorides too used *genos* and *eidos* almost interchangeably, although he gave preference to the latter. For instance, he wrote that 'there is another type (*genos*) of wild cumin, similar to the domesticated one' (*MM* 3.61), but used the word *eidos* in his description of three types of *daukos* (*MM* 3.72). Pliny, for his part, used the word '*genera* (plural of Latin *genus*)' to refer to the four types of the plant *daucus* that a certain Diodotus had distinguished (*HN* 25.110). Pliny also at times uses the Latin word *species*, subordinating it to the *genus*, as in the following example:

> There are two types (*genera*) [of ivy], as there are of other plants: the male and the female.... There are three species (*species*) in each of these types (*generum*): for the ivy is white or black and the third type is called 'helix'.
>
> (Pliny, *Natural History* 16.145 (see Chapter 5 for more detail on 'male' and 'female' plants))

However, this type of clear subordination of the species to the genus is relatively rare in Pliny, as it is in other Latin texts. Roman authors in general used the term '*genus*' (plural *genera*) much more often than the term *'species*'. The exception here is the *Alphabet of Galen*, which has a predilection for the word '*species*' (see e.g. chapter 75, p. 202 Everett; chapter 142, p. 252 Everett). In view of all these complexities, it is best to translate the Greek words *genos* and *eidos*, and the Latin words *genus* and *species* as 'type', 'kind' or 'variety', to avoid confusion with the modern notions of genus and species.

In order to define a *genos* or *eidos* of plants or animals, Aristotle and his followers used 'differences' (*diaphorai*). Aristotle first borrowed the notion of *diaphora* from Plato, but while Plato's system of classification was based on dichotomous division, Aristotle's was a complex system taking into account all parts of animals and plants. Theophrastus opened his *Enquiry into Plants* with the words 'the differences (*diaphoras*) of plants' (*HP* 1.1.1), thus indicating that his system of classification was based on the one Aristotle has devised in his *Enquiry into Animals* (see p. 75 for more detail). Dioscorides too used the word *diaphora* frequently in what appears to be a precise technical sense (see e.g. *MM* 4.82), although further research is required here. This use of *diaphorai* allowed ancient botanists to establish lists of various 'types' of a plant. For instance, Dioscorides gives four types of '*lapathon*', monk's rhubarb (*Rumex* spp.):

> Monk's rhubarb: one [type] of this plant is called *oxulapathon* [sharp monk's rhubarb]. It grows in marshes; it is tough and pointy towards the edges. Another [type] grows in gardens and is not dissimilar to the first. A third [type] is wild. It is small, similar to plantain, soft and low-lying. And there is also a fourth type (*eidos*) of the plant, which some call *oxalis* [sour wine] or *anaxuris* [trouser leg] or *lapathon agrion* [wild monk's rhubarb]. Its leaves are similar to those of the wild and small monk's rhubarb; the stem is not large, and it is pointy; the seed is red, sharp, and grows on the stem on offshoots.
>
> (Dioscorides, *Materia Medica* 2.114)

One of these types of *lapatha* is distinguished by an epithet (a descriptor), namely 'wild' (*agrion*); another has a prefix added to its name: *oxu-lapathon*, where '*oxu*' means 'sharp'; and yet another has completely different names: *oxalis* and *anaxuris*. To use epithets was by far the most common way to distinguish plants within a 'type' in the ancient world. In the following example, Theophrastus distinguishes fives types of radish by means of epithets:

> Thus the types (*genē*) of radish are: the Corinthian, that of Cleonae [in the region of Corinth], the 'smooth' from Thasos (*leiothasion*), the 'unwashed', and the Boeotian. The Corinthian is the strongest in growth, and it has an exposed root; for it pushes upwards and not downwards like the others. The 'smooth from Thasos' variety, which some call the Thracian, stands the winter weather the best. The Boeotian is the sweetest and is like a grape in shape, unlike that of Cleonae, which is long.
>
> (Theophrastus, *Enquiry into Plants* 7.4.2 (see Amigues 2003: 98 for notes on the names of these radishes))

Theophrastus identified five types of radish, four of which are qualified by a geographical epithet (Corinthian; of Cleonae; 'smooth' from Thasos; Boeotian – see Chapter 6 for more detail on these 'ethnic' vegetables). Other types of plants were distinguished by epithets indicating their colour (for instance the white and the black hellebores, see p. 72), their gender (for instance the male fern), their cultivation status (for instance the cultivated and the wild lettuce), or their habitat (for instance 'mountain nard'). From the point of view of modern botany, some of the plants distinguished by epithets in ancient texts are varieties within one species; others belong to different species within one genus; others belong to different genera within the same family; and some are different plants altogether (no relation either at the level of the genus, species or even family). Let us consider some examples.

'Ethiopian cumin' (*Aithiopikon kuminon*) occurs in numerous ancient pharmacological and culinary recipes (see Dalby 2003: 109 for references). It is relatively easily identifiable as our nigella, or black cumin, *Nigella sativa* L. (family: Ranunculaceae). The name 'Ethiopian cumin' was given to the herb by functional analogy with the cumin grown in Greece, *Cuminum cyminum* L. (family: Apiaceae): while these are completely different plants from a modern botanical point of view, they both are spices. The 'Egyptian bean' (*Aiguptios kuamos*), on the other hand, was named so by morphological analogy with the common bean (see Dalby 2003: 199–200 for references). It is identified with the pink lotus, *Nelumbo nucifera* Gaertn. (family: Nelumbonaceae), whose seed resembles a bean. Similarly, the 'sea cabbage' (*krambē hē thalassia*), and numerous other sea plants, were named by morphological analogy with the common cabbage, even though Dioscorides noted that it was 'completely different from the cultivated cabbage' (*MM* 2.122). Note that the names 'black cumin' and 'sea cabbage/cole' are still in use in vernacular English. We shall call these vernacular binomials 'pseudo-binomials' to avoid confusion with the scientific Linnaean binomials. Other examples of pseudo-binomials in English would be 'Jerusalem artichoke' or the 'Lily of the valley' (see Greene 1909: 124).

In all three examples examined thus far (Ethiopian cumin, Egyptian bean and sea cabbage), the ancients were well aware that they were dealing with different types of plants. They knew that dropping the epithets (Egyptian, Ethiopian, sea) would have caused much confusion. Such confusion did sometimes arise with homonymous plants mentioned without their epithet, as in the case of the two *helleboroi* ('hellebores'), of which Theophrastus writes:

> Black and white hellebores are in some ways homonyms (*homōnumoi*), but people disagree as to their appearance. For some, they are similar, with the exception that their roots differ in colour: white for the one, black for the other. But others say that the leaf of black hellebore is like that of bay, that of white hellebore is like that of leek, and that their roots are similar, except for their colours.
>
> (Theophrastus, *Enquiry into Plants* 9.10.1)

Theophrastus knew that there were two types of hellebore but could not rely on his sources to give him a clear account of their differences. His informants here were the drug-sellers (see Chapter 2). The philosopher recounts how the drug-seller Thrasyas publicly consumed several roots of 'hellebore' without being purged, thus creating wonder among his audience – 'hellebores' were used in ancient pharmacology for their strong purgative properties. Thrasyas, however, was beaten at his own game, when a shepherd presented himself to the drug-seller and consumed a whole bundle of the root, thus destroying Thrasyas' reputation (*HP* 9.17.1). Ancient medicinal recipes often recommended the use of 'hellebore' but rarely specified whether the white or the black was needed (see Lloyd 1983: 126–127). Scholars have identified the black hellebore with *Helleborus cyclophyllus* Boiss. (family: Ranunculaceae) and the white one with *Veratrum album* L. (family: Melanthiaceae).[8] The two plants are completely different from the point of view of modern botany, and can clearly be differentiated by their appearance. It seems ancient pharmacologists had little interest in divulging all their knowledge regarding these two plants, since both had the same pharmacological effect. But while that situation might have been acceptable for a patient in need of a purge, or for a physician purchasing a dried plant from the market, it was not for a botanist like Theophrastus.

While in the case of 'hellebore' the ancients tended to drop the epithet, in others they considered the main plant name to be redundant, as in the case of 'apples'. The name 'apple' (Greek: *mēl(e)a*/Latin: *mala*) was applied to numerous trees bearing fleshy fruits: the common apple, the peach, the apricot, the citron, the quince, the pomegranate, etc. (see Dalby 2003: 19–20, 88, 252, 266, 275–276). To avoid confusion, epithets were used: 'Persian apple' (= our peach); 'Armenian apple' (= our apricot); 'Median apple' (= our citron); 'Cydonian/Cotonean apple' (= our quince); 'Punic apple' (= our pomegranate, but note that this appellation is only used in Latin, in Greek the pomegranate is *rhoa*). Several of these fruits came to be designed by their geographical epithet only: *kudonia*/*cotonea* (the quince); *punica* (the pomegranate); *persica* (the peach).[9] From the Linnaean point of view, the 'Persian apple' (*Prunus persica* Stokes, family: Rosaceae) and 'Armenian apple' (*Prunus armeniaca* L., family: Rosaceae) are part of the same genus, that of the *Prunus*; but the Greek 'Median apple' (*Citrus medica* L., family: Rutaceae) is part of the *Citrus* genus; the Greek Cydonian apple (*Cydonia oblonga* Mill., family: Rosaceae) is the sole member of the genus *Cydonia*; and the Latin 'Punic apple' (*Punica granatum* L., family: Lythraceae) is part of the small *Punica* genus (which also includes the Socotra pomegranate, *Punica protopunica* Balf.f.). Within the apples *stricto sensu*, the Cydonian ones, the Persian ones, etc., the ancients distinguished multiples varieties, also by means of epithets. For instance, Pliny lists four types of 'Persian apples' (= peaches): the hard one (*duracina* = our nectarine), the Gallic (*Gallica*), the Asiatic (*Asiatica*), and the *supernatia* (*HN* 15.39–40, see also Palladius, *OA* 12.7.3; Columella, *RR* 10.410–412).

Ancient authors in general, and Latin ones in particular, had a predilection for enumerating varieties of common fruits and vegetables. For instance Theophrastus listed three varieties of cabbages (*HP* 7.4.4), as did Cato (*RR* 157.2); Nicander enumerated four (fragment 85 = Athenaeus 9.370a); Pliny five (*HN* 20.79–96); and Columella in his poem had fifteen (10.178–193). Pliny also tended to establish a hierarchy in his lists, as he did in his long enumeration of vines (*HN* 14.21–53, see Kircher-Durand 1993), which starts with the Aminean vine 'to which the highest rank should be awarded on account of the body of that wine and its life, which undoubtedly improves with age'.

The issue of ancient plant *eidē* and *genē* is complicated by the fact that the ancients thought no *eidos* or *genos* was fixed; they believed that several plants could change into others (see French 1994: 98–99). Theophrastus discussed various such changes in book two of the *Enquiry*, distinguishing between those changes that are contrary to nature, and therefore of interest to the soothsayers (acid pomegranate producing sweet fruit; wild fig turning into a cultivated one; cultivated olive turning into wild one, and conversely), and those that are ordinary:

> Among other plants, it seems that catmint [i.e. wild mint] turns into mint [i.e. cultivated mint], unless it is mastered by special care, and for that reason they often transplant it; and similarly wheat turns into darnel. Now in trees these changes, if they occur, are spontaneous, but in annual plants they occur through deliberate practice. For instance, einkorn wheat and emmer wheat change into wheat if bruised before sowing; and this does not happen immediately but in the third year. This is comparable to the changes in seeds according to regions.
> (Theophrastus, *Enquiry into Plants* 2.4.1 (see also *HP* 8.7.1; *CP* 4.4.5; 4.5.4; 5.3.7))

These examples are often repeated in ancient literature. To them are added, among others, the case of the white poplar turning into a black poplar with age (*CP* 2.16.2; Nicolaus, *On Plants* 1.7.123, 821a28–29); the old cabbage seed turning into turnip (Palladius, *OA* 3.24.7); the basil changing into thyme or mint (*CP* 5.7.2; Gargilius Martialis, as quoted in Palladius, *OA* 5.3.4; Nicolaus, *On Plants* 1.7.123, 821a31); and the larch turning into torch pine through the effect of disease (Pliny, *HN* 16.44). Galen reported at length an experiment that his father carried out to determine whether darnel originated from wheat and barley:

> My father, having become fond of farming in his advanced age, once sowed wheat and barley, after painstakingly picking out from them every single seed of another type (*heterogenes*) mixed with them, in order to understand clearly whether the generation of darnel and aegilops occur as a result of a change from them [wheat and barley], or whether these seeds have their own nature. As it happens, a non-negligible amount of darnel was generated among the pure seeds of wheat, but few among the barley; instead, there was plenty of aegilops.
> (Galen, *Properties of Foodstuffs* 1.37, 6.552 Kühn)

Galen went on to recount other experiments of his father.[10] Now, from the point of view of modern botany, it is possible for a cultivated plant to return to wild state, but is not possible for wheat (*Triticum* spp. family: Poaceae) or barley (*Hordeum vulgare* L., family: Poaceae) to turn into darnel (*Lolium temulentum* L., family: Poaceae), or cabbage (*Brassica oleracea* L., family: Brassicaceae) to turnip (*Brassica rapa* L., family: Brassicaceae). But clearly, ancient definitions of *genos*/*eidos* are different from modern ones.

In this section on *eidos* and *genos*, we have encountered various *diaphorai* used to classify plants in the ancient world – we will return to several of these in the remainder of this book. We have not, however, discussed the most important *diaphora*: that found in parts of plants, which led Theophrastus to his division of plants into four categories (trees, shrubs, undershrubs and plants), a division that proved extremely influential throughout antiquity and beyond. We turn to it.

3.4 Ancient systems of plant classification

3.4.1 Theophrastus' classification and the parts of plants

As stated earlier, Theophrastus opens his *Enquiry* with the words 'the differences of plants'. Throughout book one, Theophrastus enumerates various *diaphorai*: differences in cultivation status (cultivated or wild); in habitat (earth or water); in fertility (barren or fruitful); whether the plant is evergreen or deciduous, flowering or not. These various *diaphorai* also make appearances in the remainder of the *Enquiry*, but they are subsumed to the *diaphora* in plant parts.[11] In putting emphasis on the parts of plants, Theophrastus was following Aristotle, who had said that 'for animals, the matter (*hulē*) is in their parts' (*GA* 1.1, 715a10). This was, however, difficult. On the one hand, plants were less differentiated with regards to parts than animals. Thus, in *Causes of Plant Phaenomena*, Theophrastus asserted that 'plants are constituted more of undifferentiated (*homoiomerē*) parts than animals' (*CP* 5.2.1), echoing the words of Aristotle who had stated that 'the nature of plants, being immobile, is not varied (*polueidēs*) with regards to differentiated (*anomoiomerōn*) parts' (*Parts of Animals* 2.10, 656a1–2). On the other hand, it is difficult to determine what counts as a 'part' in a plant:

> Now the differences in their modes of generation (*genesin*), properties (*pathē*), and modes of life are comparatively easy to observe and are simpler, while those difference in parts present more variety. For indeed it has not even been determined what one should call, and what one should not call, 'parts', and this involves some difficulty.
>
> (Theophrastus, *Enquiry into Plants* 1.1.1)

Indeed, in the Aristotelian scheme, parts ought to be permanent (as they are in animals), but some parts (e.g. leaf, fruit) plainly are not in plants that go through annual cycles. As pointed out earlier, however, Theophrastus, although he often made use of analogies between plants and animals, warned against the dangers of over-zealous comparisons. The question of parts of animals and plants was no exception here:

> When considering the differences between plants and their distinctive nature, one must take into account their parts, their afflictions, their modes of generation, and their modes of life. For plants are not endowed with characters and activities.... But perhaps we should not seek complete correspondence [between plants and animals] in respect to those things linked to generation any more than in other respects, and we should count as 'parts' (*merē*) even those things the plants beget, such as their fruits (*karpous*), even though we [do not consider] embryos [to be part] of animals.
>
> (Theophrastus, *Enquiry into Plants* 1.1.1–3)

Thus within the first few paragraphs of the *Enquiry*, Theophrastus situated himself within the Aristotelian tradition while simultaneously breaking free from it. He then described three types of differences in the parts of plants:

> The *diaphorai* in parts, to take a rather schematic view, are three: either the presence or absence of parts, as for instance the leaves and fruits; or they are not similar or equal; or thirdly, they are not [disposed] similarly.
>
> (Theophrastus, *Enquiry into Plants* 1.1.6)

These three types of differences in parts led Theophrastus to classify plants into four main categories: the tree, the shrub, the undershrub and the herb. These are succinctly described in the following passage:

> The first and most important classes, those which comprise all or nearly all plants, are tree, shrub, undershrub, and herb. A tree is a thing which springs from the root with a single stem, having nodes and several branches, and it cannot easily be uprooted; for instance, olive, fig, and vine. A shrub is a thing which rises from the root with many branches; for instance, bramble and Christ's thorn. An undershrub is a thing which rises from the root with many stems as well as many branches; for instance, cabbage and rue. A herb is a thing which comes up from the root with its leaves and has no main stem, and the seed is borne on the stem; for instance, corn (*sitos*) and vegetables (*lachana*).
>
> (Theophrastus, *Enquiry into Plants* 1.3.1)

We will return to individual parts of plants in Chapter 4, concentrating here instead on the principles of classification based on plant parts. The key passage quoted above calls for several comments. First, Theophrastus used all three categories named in the passage of the comedian Epicrates cited at the beginning of this chapter: vegetables, herbs and trees. 'Tree' and 'herb' are two of Theophrastus' four main categories, while 'vegetable' is a sub-category of the 'herbs'. Second, Theophrastus chose as names for his categories words that already existed in the Greek vocabulary, but gave them a technical scientific definition.

Thus, before Theophrastus' time, the word *phruganon* designated dry wood, twigs gathered for fuel (see Greene 1909: 108). Third, some of Theophrastus' categories were very large, and would need further division, as he did in this passage with the 'herb' category which is divided into 'cereals' and 'vegetables'. Fourth, the boundary between these categories can seem a bit blurry. Theophrastus addressed this point, stating that 'it is necessary to accept and take these definitions (literally: boundaries) as being general and as applying on the whole' (*HP* 1.3.2). In general, the philosopher rejected all systems of classification that are too rigid, arguing that classification can only be 'typical' (*HP* 1.3.5). Throughout the *Enquiry*, Theophrastus sometimes refers to classifications he has inherited from his informants: classification of trees by the use that can be made of their wood (*HP* 5.7), a classification which he may have borrowed from woodworkers; classification of the pines into 'Idaean pine' and 'maritime pine' by the people of the Ida (*HP* 9.2.5).

This division into parts is the main structuring tool in *Enquiry into Plants*, starting in the middle of book two.

Book 2b: Domestic trees and shrubs
Book 3: Wild trees and shrubs
Book 4: Trees and shrubs considered from an environmental point of view
Book 5: Trees and shrubs and their timber
Book 6: Undershrubs
Book 7: Herbs
Book 8: Cereals
Book 9: Medicinal herbs.

The 'tree' and 'shrub' categories cover three and a half books (books two to five), and that on herbaceous plants covers three books (books seven to nine). The 'undershrub' category, on the other hand, occupies only one book: book six. Within each category, one of the main subdivisions is that based on the environment, between wild and domesticated, a division we will study in detail in Chapter 6.

In book six, in addition to considering the wild or cultivated status of undershrubs, Theophrastus divided them into inerm (without thorns, gr. *anakantha*) and thorny (*akanthōdē*). From chapter six of that book, Theophrastus turned to plants used to make wreaths, which fall between the 'undershrub' category and the 'herb' category:

> We must attempt to discuss the plants used in wreaths (*stephanōmatōn*) in general so that the whole type (*genos*) may be taken into consideration. For the nature (*phusis*) of plants used in wreaths has a somewhat peculiar position since it mixes together undershrubs and herbs.
>
> (Theophrastus, *Enquiry into Plants* 6.6.1)

He went on to include both undershrubs and herbs in this 'wreath' category, which is a category based on usage:

> They can be divided into two groups according to their usage (*chreian*). For some of them have a useful flower; and among these some have sweet-smelling flowers, like the violet, some without smell, like the carnation or the wall-flower. For others it is the branches, leaves, and the entire growth that are sweet-smelling, as in the case of the tufted thyme, catmint, bergamot mint and others.
>
> (Theophrastus, *Enquiry into Plants* 6.6.2)

Wreaths played an important function in ancient societies: they were used in numerous religious rituals; worn as prizes by athletes and military victors; and worn at symposia and other dinner parties (see Campbell 2012). Theophrastus did not indicate his source for this classification by usage, but it is possible that he had met with wreath-makers. Both Pliny and Dioscorides mention these people as their sources (Dioscorides, *MM* 3.75; Pliny, *HN* 21.68, see Chapter 2). In any case, it is important to stress Theophrastus' recourse to a category based on usage rather than morphology, and straddling his 'undershrub' and 'herb' categories. Here is also the place to mention a long fragment of Nicander's *Georgics*, in which the poet discusses various plants suitable for wreaths: gillyflower, rose, ivy, lily, iris, poppy, marjoram, ferns, acanthus, etc. (fragment 74 = Athenaeus 15.683a–684f).

Herbaceous plants are extremely numerous and Theophrastus acknowledged the need for some further subdivision of that category. He settled on a main sub-division into cereals and 'vegetables' (the translation 'pot-herbs' is sometimes found):

> Let us carry on our discussion with the herbaceous plants (*poiōdōn*). For it is the last of the types (*genōn*) we distinguished at the start, the one which includes more or less all the vegetable type (*lachanēron*) and the cereal one (*sitōdes*). And first we must start by talking about the vegetables (*lachanōdous*), starting with the cultivated ones (*hēmerōn*), since it happens that they are better known than the wild.
>
> (Theophrastus, *Enquiry into Plants* 7.1.1)

'Vegetable' is a category which Theophrastus himself found problematic, as becomes clear at the beginning of his discussion of the wild vegetables:

> We must now speak of the so-called 'uncultivated (*arouraiōn*) plants', and in general, of those herbaceous plants that are not edible. For we call 'vegetables' (*lachana*) those of which we make use, but in general, [the class] also includes those other plants – hence we must talk about those too. Are also called 'vegetables' the chicory, dandelion, *chondrilla*, cat's

> ear, groundsel, and in general those that are called 'chicory-like' on account of the similarity of their leaves.
>
> (Theophrastus, *Enquiry into Plants* 7.7.1)

The Greek word '*lachanon*' primarily refers to plants that are eaten in accompaniment to cereals or legumes; most of these plants were cultivated – grown by agricultural techniques – but not all of them. Thus the catalogue of foods included in the Hippocratic treatise *On Regimen* II includes the following plants under the '*lachanon*' category (chapter 54, Loeb 4.328–332 Jones = 6.556–562 Littré): garlic, onion, leek, radish, cress, mustard, rocket, coriander, lettuce, anise, celery, basil, rue, asparagus, sage, nightshade, purslane, nettles, catmint, mint, sorrel, orach, blite, cabbage, beet, pumpkin, turnip, pennyroyal, marjoram, savory, thyme, hyssop, and various other wild vegetables (note the presence of the pumpkin here). While the 'vegetable' category works relatively well from a dietetic point of view, it is not as successful in a morphological classification like that of Theophrastus. Despite the difficulties inherent to that category, Theophrastus devoted book seven of the *Enquiry into Plants* to vegetables. He probably could not do without a category that already had a long history. In book eight, he dealt with cereals, leguminous plants and plants which, like millet and sesame, lack 'a common designation' (*HP* 8.1.1). Book nine, again, makes use of divisions that are based on usage: there Theophrastus covered aromatic plants (chapters 1–7) and plants called 'roots', that is, medicinal plants (chapter 8–19).

Within the *Enquiry into Plants*, and more particularly within books six to eight, other smaller groupings can be observed. They are designated in Greek by epithets in -ōdēs: the bulbous plants (*bolbōdēs*: e.g. *HP* 7.2.2); the fennel-like plants (*narthēkōdēs*: e.g. *HP* 6.1.4); the thorny plants (*akanthōdēs*, see p. 77); the chick-pea-like plants (*kichoriōdēs*: e.g. *HP* 7.7.3); the vine-like plants (*botruōdes*, e.g. *HP* 3.12.7); etc. Several of these Greek epithets in -ōdēs have survived in modern botanical Latin, as name of families such as Chicoriaceae or Narthеciaceae.

Theophrastus' system of classification proved very influential. Nicolaus' system is perhaps the closest to Theophrastus', with some slight differences worth pointing out:

> A tree is that which has a stem growing from its root, from which numerous branches ramify, such as the olive tree and fig tree. As to the plant which is midway between tree <and herb> from which we have said that it is a thamnos or a shrub, it has many branches growing from the root, like the plant called paliuros, and like cane and bramble. Vegetables are those plants which have many stems growing from the root with many branches, such as rue and cabbage. Herbs are those which bear leaves from the root and have no stems … Some plants tend towards two sides, like the vegetable called mallow, since it is both a herb and a vegetable, and likewise the beetroot.
>
> (Nicolaus, *On Plants* 1.4.89–91, 819b4–19; translation: Drossaart Lulofs and Poortman 1989: 154)

Nicolaus, like Theophrastus, has four categories, and uses the same – or very similar – examples to Theophrastus to illustrate each one. For the 'shrub' category, the Arabic translator used the Greek word '*thamnos*'. For the 'undershrub' category, on the other hand, he used a word that is best translated in English as 'vegetable'. Again, the 'vegetable' category is problematic as some plants seem to fall both in that category and the 'herb' category. Other authors borrowed Theophrastus' classification system, but adapted it to their own purposes and/or subsumed it to another principle, as we will now see.

3.4.2 Dioscorides' classification by powers of plants

Dioscorides referred to all four of Theophrastus' categories (tree, shrub, undershrub and herb) (see Touwaide 1999a: 234). He sometimes stated at the beginning of his entry on a plant that it is a tree (*dendron*, see e.g. *MM* 1.104: 'the willow is a well-known tree'), a shrub (*thamnos*, see e.g. *MM* 1.20: 'camel's thorn … is a woody shrub'), a herb (*poa*, see e.g. *MM* 1.29: 'another elecampane … it is a herbs'; *botanē*, see e.g. *MM* 3.44: 'the sowbread: it is a sweet-smelling herb'). More research is needed to determine whether there is a significant difference between Dioscorides' *poai* and *botanai*, two terms that can be translated as 'herb'. Dioscorides never said that a plant was an undershrub, but rather stated that plants are 'undershrub-like' (*phruganōdēs* or *phruganoeidēs*, see e.g. *MM* 3.54: 'Cretan hartwort … it is an undershrub-like little herb') or uses the diminutive 'small undershrub' (*phruganion*, see e.g. *MM* 3.91: 'Madwort: it is a small undershrub'). He used two diminutives of *thamnos*: *thamnion* (e.g. *MM* 2.162: 'The sneezewort is a little shrub') and *thamniskion* (e.g. *MM* 3.36: 'Thyme … it is small shrub that is undershrub-like', *phruganoeidēs*). For plants that fell between two categories he used epithets such as 'shrubby' (*thamnōdēs* or *thamnoeidēs*, e.g. *MM* 1.88: 'the tree heath is a shrubby tree'); 'tree-like' (*dendrōdēs* or *dendroeidēs*, e.g. *MM* 1.103: 'The chaste tree is a tree-like shrub'); or 'undershrub-like' (see above for example).

In addition to Theophrastus' four categories, Dioscorides also referred to plants as 'vegetables' (*lachanon*; see e.g. *MM* 2.116: 'the charlock is a wild vegetable') and stalks (*kaulos*, diminutive *kaulion*, e.g. *MM* 2.144: 'The starflower is a tender little stalk'). Finally, like Theophrastus, Dioscorides used numerous epithets ending in -ōdēs or -oieidēs to group plants in categories such as the 'thorny' (*akanthōdēs*, e.g. *MM* 1.100: 'Dyer's buckthorn … is a thorny plant') or the 'fennel-like' (*narthēkoeidēs*, e.g. *MM* 3.82.1: 'Spurge: it is a Libyan fennel-like tree').

While Dioscorides made use of Theophrastus' classification, he only employed it as a subordinate to his main classification, one based on the effects of *materia medica*, which he expounded in his preface: 'here a different arrangement shall be used, one in which each type of plant is listed according to its powers (*dunameis*)' (*MM*, preface 5). By '*dunamis*', Dioscorides meant the physiological effects of the plants on the human body (and sometimes on the animal body, as Dioscorides took into account veterinary uses of *materia medica*). This was a complex system, which was not necessarily based on the most easily observable effects. For instance, not all of Dioscorides' purgative plants were classified together, as Lily Beck's medical index makes very clear (2005: 499–540, p. 530 for purgatives). As noted by Riddle, who has studied this system in detail, it was so sophisticated that '[Dioscorides] actually classified some plants by their active chemical constituents. For example, he grouped together mydriatic alkaloids (atropine, scopolamine), some volatile oils in the Umbelliferae, some alkylating agents with cytotoxic compounds, and acetylsalicylic acids' (1985: 97; see also Thomson 1955: 23; Touwaide 2007: 43).

Unfortunately, Dioscorides never fully explained his system, merely hinting at it in the prefaces to the five books of the *Materia Medica*, as is the case here:

> Dearest Arius, in the previous book, which was the first I composed on *materia medica*, I dealt with the spices, oil, perfumes, and trees, as well as their saps, tears, and fruits. In this book, which is the second, I will discuss animals, honey, milk, animal fats, and the so-called cereals, as well as vegetables, adding to them those herbs that are used for their sharp property, on account of their similarity, such as garlic, onion, and mustard, in order to group together products with similar effects.
>
> (Dioscorides, *Materia Medica*, book 2, preface)

Here Dioscorides explicitly stated that various sharp plants have similar effects on the body. The anthropologist Scott Atran has noted that such groupings according to properties or virtues, so-called 'unnatural groupings', are actually often based on morphological observation. Here is what he writes about Theophrastus' classification in book nine – his conclusions are even more applicable to the writings of Dioscorides:

> A closer examination of Theophrastus' *Historia plantarum* ... shows that a discussion of cultural virtues is predicated on the following premises: that shared virtues are necessary consequences of similarity in underlying essential nature (*physis*), and that there is often no better sign of similarity in nature than a popular intuition of overall resemblance in readily visible morphological aspect ... In other words, learned communication *presupposes* the *principled* acceptance of morphological types ... Similar considerations apply to the folkbiology of any society. There is, it appears, a (universal) presumption to the effect that visible organic types have underlying natures.
>
> (Atran 1987: 197)

As we indicated in Chapter 1, a division between 'pure' and 'applied' botany is rather meaningless in the context of ancient botany, and Dioscorides' classification, which in modern terms we could call 'pharmacological', is a good case in point: it is actually based on observation that goes well beyond the purely 'pharmacological' and encompasses the morphology and physiology of plants.

Dioscorides was critical of his predecessors' classificatory principles, and in particular of the alphabetical classification (*MM*, preface 3). His followers, however, did not always perceive the quality of Dioscorides' system and alphabetised the material they borrowed from *Materia Medica* (see p. 83 for Galen). Dioscorides' *Materia Medica* itself was alphabetised in late antiquity. Thus, books eleven, twelve and thirteen of Oribasius' *Medical Collection* are an alphabetical summary of Dioscorides' *Materia Medica.* Whether Oribasius himself alphabetised Dioscorides' material or had access to an already-alphabetised version is difficult to determine (see Barbaud 1994, on the alphabetical order in antiquity, see Daly 1967). The famous Vienna Dioscorides manuscript is also organised alphabetically. Further research would be required to determine how works derived from Dioscorides, such as *Medicines made from Female Herbs* are organised, as it is certainly not presented in alphabetical order. We now turn to Galen, who used the alphabetical system in some of his writings, but appears to have been aware of other systems.

3.4.3 Classification of plants in the works of Galen and other medical writers

As a learned physician, well versed in the works of Theophrastus, Galen sometimes appears to follow the philosopher's classification:

> I advise the physician to have experience (*empeiron*), if possible, of all plants, but if not of at least the majority of them, and certainly of those that are most used. Their genres (*genē*), or if you prefer to call them 'differences' (*diaphoras*) are as follows: the trees (*dendra*), the shrubs (*thamnoi*), the herbs (*poai*), the thorns (*akanthai*) and the undershrubs (*phrugana*).
>
> (Galen, *On Antidotes* 1.5, 14.30 Kühn)

Here Galen mentioned all four of Theophrastus' classes, placing the undershrubs at the end of the list (the word *phruganon* appears very little in the Galenic corpus), and adding to them the 'thorn' category, perhaps because many thorny plants are used in medicine. Like Dioscorides, Galen in *Simples* sometimes opened a paragraph on a plant by stating that it is a tree (*dendron*, e.g. *Simples* 7.10.19, 12.22 Kühn: 'The cherry is a tree that bears fruit'); shrub (*thamnos*, e.g. *Simples* 6.1.22, 11.821 Kühn '*Alimon*: this plant is a shrub'; *thamniskos*: e.g. *Simples* 6.1.68, 11.841 Kühn: 'Milk vetch is a very small shrub'); herb (*botanē*: e.g. *Simples* 7.15.17, 12.92 Kühn: 'Orchid: the same herb is also called dog's testicle'; *poa*: e.g. Simples 6.1.42, 11.830 Kühn: 'Sea navel is a bitter and sharp herb'); a thorn (*akantha*, e.g. *Simples* 6.1.72, 11.842 Kühn: 'spindle-thistle ... is one of the thorns'); or a vegetable (*lachanon*, e.g. *Simples* 6.2.8, 11.851 Kühn:

'Blite is an edible vegetable'). Like Theophrastus and Dioscorides, Galen also used numerous epithets in -ōdēs to refer to smaller groupings of plants. For Galen, however, Theophrastus' morphological classification of plants was subordinate to a classification by powers in his *Simples*, and to a classification by dietetic use in *Foodstuffs*. Let us start with the former.

Galen devised a system, in theory simple, to classify drugs according to their effects on the body. He identified various powers, four of which were most prominent: heating, cooling, drying and moistening. These powers could manifest themselves in four degrees of intensity: weak, obvious, strong and massive (see Vogt 2009 for an introduction to this system). For instance, Galen writes of *elaphoboskon* (perhaps parsnip): 'Concerning parsnip: Parsnip is warming and thinning in properties, and it follows that it is drying in the second degree, especially in cases of inflammations' (Galen *Simples* 6.5.6, 11.873 Kühn). However, in *Simples*, Galen did not present drugs according to their powers (for instance, all cooling drugs together), opting instead for the alphabetical order. In the preface to book six of *Simples* (11.792 Kühn), Galen admitted that the otherwise unreliable Pamphilus (see Chapter 2) was his model for the alphabetical (*kata stoicheion*) structuring principle (see Flemming 2007: 254–255). As its title indicates, the Latin *Alphabet of Galen* is also organised alphabetically, although as noted in Chapter 1, this work is certainly not by Galen, and bears more similarity to Dioscorides' work than to Galen's *Simples*.

In one of his dietetic treatises, *Properties of Foodstuffs*, Galen divided foods into cereals and legumes (book one); vegetables and fruits (book two); and animal products (book three). Galen took care to define his categories. Here is what he had to say about the 'fruit' category, and how 'fruits' differ from 'seeds'. We have given in brackets the Greek words for 'fruit' and 'seed'. Note that the Greek language has declensions, which means that the same word has different endings according to its case (its grammatical function); thus *sperma*, *spermatos* and *spermata* are all forms of the same word:

> The foods (*sitia*) in the first books were all seeds (*spermata*) of plants. Now I will start with the fruits (*karpōn*), after having distinguished their seeds (*spermata*), particularly because many think that the fruit (*karpon*) is no different from the seed (*sperma*). Now the seeds (*spermata*) already discussed differ slightly from fruits (*karpōn*), but those of which I will now talk are very different. For the fig is the fruit (*karpos*) of the fig tree, but the seed (*kenchramis*) within the fruit is the seed (*sperma*) of the fig. In the same way, while the entire berry (*rhagos*) is the fruit (*karpou*) of the vine, the grape-stone (*gigarton*) alone is the seed (*sperma*) of the vine. Similarly, the fruit (*karpos*) of the pear tree and that of the apple tree are the pear and the apple, but the seed (*sperma*) is the three or four pips (*kokkoi*) in their middle. What to say of the pumpkins and cucumbers, both the 'ripe' types and the others, and all the other [plants] of that sort? For in these the entire fruit

> (*karpos*) is different from the seed (*spermatos*).... There are also the fruits (*karpoi*) of the trees and from cultivated vegetables (*lachanōn*), among which some physicians include gourds, melons, and pumpkins, from which I shall start in this book.
>
> (Galen, *Properties of Foodstuffs* 2.1, 6.556–558 Kühn)

Galen is here talking about 'fruits' as a dietetic category, which he takes care to define as distinct from seeds. Seeds can take various forms, to which various colloquial names are given: seedlets, grape-stones and pips. To Galen, the 'fruit' (*karpos*) is the edible fleshy part rather than the seed. Galen was aware that his understanding of the words 'seed' (*sperma*) and 'fruit' (*karpos*) was different from that of, for instance, Theophrastus, who studied plants from a morphological point of view rather than a dietetic one. Similarly, Galen's definition of the 'vegetable' is dietetic rather than morphological: a vegetable is an edible small plant (see *Foodstuffs* 2.39, 6.624 Kühn). As we saw in Chapter 2, various medical authors wrote now-lost treatises on vegetables; that category was important from a dietetic point of view even though it posed many morphological issues.

A similar classification of foodstuffs is to be found in all dietetic treatises of antiquity, both anterior and posterior to Galen. Thus the catalogue of foods in the Hippocratic *Regimen* II is organised as follows: cereals and cereal products (chapters 40–44); legumes and seeds (chapter 45; note that no general name is given to all these products); animals and animal products (chapter 46–51); water and honey (chapters 52–53); vegetables (*lachana*, chapter 54); and seasonal fruits (*opōrē*, chapter 55). Much later in antiquity, Paul of Aegina organised the dietetic section of his book one as follows: herbs and vegetables (1.74), shoots (*asparagoi*, 1.75), plants with edible roots (1.76), truffles and mushrooms (1.77), cereals (1.78), pulses (1.79), seasonal fruits (1.80), fruits from trees (1.81), meats and animal products, etc. (1.82 and following) (CMG 9.1, pp. 53–60 Heiberg). The classification of plants within dietetic treatises is also similar to that found in ancient agronomical treatises, which we will consider together with Pliny's encyclopaedic writings.

3.4.4 Classification of plants in the works of Pliny and the agronomical writers

Pliny's classification of plants has come under criticism for being rather haphazard. For instance, Guy Ducourthial writes: 'Pliny did not take care to elaborate a coherent and rational system to classify the various species of plants listed. Thus, the order of exposition he adopts to present these plants relies on a very loose system of classification' (2003: 63, our translation from the French).[12] We would argue that this judgement is far too harsh: Pliny in fact uses various systems of classification in books twelve to twenty-seven of the *Natural History*. Books twelve to sixteen are devoted to trees and shrubs that grow wild and are organised geographically. Books seventeen to nineteen are what one could call

Pliny's agronomical books, where he focuses on the plants that are most important in farm management. Books twenty to twenty-three are what we could call Pliny's dietetic treatises, where he focuses on plants as foods. Books twenty-four to twenty-seven are the pharmacological books.

Pliny starts his discussion of trees in book twelve with the following words:

> The riches of the earth remained for a long time hidden, and the trees and forests were thought to be the supreme gift offered to humans. These first provided them with food; their foliage softened their caves; their bark served them as clothing; and even now there are still peoples who live in this way.... For this reason, it is good to follow the order of life and to speak first of trees before other [plants], and to bring forward origins for our customs.
>
> (Pliny, *Natural History* 12.1–2)

Pliny's reasons for starting with trees were, therefore, different from those of Theophrastus. He was not interested in morphology, but rather in a form of ethnography: trees are what primitive people turn to for their shelter and food. Book twelve is then devoted mostly to trees and shrubs (*frutex*) that are exotic, that is, that cannot be grown in Italy, such as frankincense or cinnamon.[13] Book thirteen opens with an excursus on perfumes, since so many of the trees presented in book twelve produced scented oils; it then turns to other exotic trees and shrubs, starting with the palm (*HN* 13.26). The paragraph on the papyrus, a marsh shrub (*frutex*, *HN* 13.68) opens another excursus on writing. The book ends with considerations on sea 'trees' (*arbores*) and 'shrubs' (*frutices*) (*HN* 13.135–142). It is important to note that all the exotic trees listed in books twelve and thirteen come from regions east and/or south of Italy. With book fourteen, Pliny turned his attention to trees that grow in Italy, as well as in other regions: book fourteen focuses on the vine and wine; book fifteen on olive trees and oil, other trees producing oleaginous fruits and nuts, and fruit trees (apple, pear, peach, etc.), the myrtle and the laurel. In book sixteen, Pliny discusses forest trees, noting that some of the biggest forests grow in Germany, that is, north of Italy (*HN* 16.5). He also includes in that book other plants that like cold conditions, including some water shrubs (*HN* 16.156–180), and plants that grow on trees, such as mistletoe (*HN* 16.245–251). In this book, the encyclopaedist also draws a distinction between plants that shed their leaves and those that do not, between deciduous and evergreen (see in particular *HN* 16.78 and 82). To sum up, books twelve to sixteen of the *Natural History* are organised geographically, starting with 'hot' regions and ending with 'cold' ones. They focus on trees and shrubs, even though there are various excursus.

Pliny opened book seventeen with the following statement:

> We have now talked about the nature of trees that grow of their own accord on the earth and in the sea; it remains to discuss those that in

> truth, rather than being born, occur through skill (*arte*) and human devices.
>
> (Pliny, *Natural History* 17.1)

In reality, Pliny had discussed many cultivated trees in books twelve to sixteen, but in book seventeen the focus very much turned to agriculture, with consideration given to the most beneficial soils for plants, methods of planting, grafting techniques, advice on fertilisation, diseases affecting crops, etc. Book seventeen focuses on fruit trees and the vine. Book eighteen deals with cereals and legumes (*legumina*), and ends with a long calendar of farm activities with astronomical and meteorological considerations (*HN* 18.201–364). Book nineteen opens with the cultivation of plants used in textile-making (flax, esparto) and dyeing, followed by the cultivation of kitchen gardens (*cura hortorum*, starting at *HN* 19.49). Book nineteen also includes an excursus on truffles and silphium (*HN* 19.33–46). A similar organisation to that of Pliny's books seventeen to nineteen is to be found in other agronomical treatises. For instance, Columella discusses cereals and legumes in book two of *On Agriculture*; he deals with the vine in books three and four and a part of book five; and turns to the olive and other fruit trees in book five. Isidore of Seville, for his part, organised plants within his agronomical book (book seventeen), as follows: cereals, legumes, vine, trees, aromatic trees, herbs, garden vegetables, aromatic garden vegetables.

With book twenty, Pliny turned his attention to plants as foods, stressing their dietetic properties. He opened the topic in the following way:

> From this point, we begin our discussion of a most important work of nature, that is, the proper foods of people, and we shall urge them to acknowledge that their means of living are unknown to them.... I start with garden plants (*hortensiis*).
>
> (Pliny, *Natural History* 20.1–2)

Book twenty is devoted to the properties of kitchen-garden plants. Book twenty-one deals with wreath flowers; plants used as spices and in perfumes; bees and honeys (*HN* 21.70–86); and ends with edible plants that grow wild (*HN* 21.87–184). More research is needed on the organisation of books twenty and twenty-one, but it seems to us that it is very similar to that found in ancient dietetic treatises. The organisation of book twenty-two is – admittedly – more difficult to discern, but the book is on the whole devoted to the properties of foreign plants, that is, plants that do not grow in Italy. In book twenty-three, Pliny focuses on the dietetic properties of fruits and their products, including vine and wine, and olives and oil. Again, the organisation in book twenty-three is a traditional one that would have probably been used in treatises on the medicinal properties of fruits.

In book twenty-four, Pliny shifts the emphasis once again, now focusing on the pharmacological (rather than dietetic) properties of plants. Book twenty-four deals with medicines from trees, shrubs and thorns. Here too more research is needed as to the detailed organisation, but it is to be noted that some alphabetical sequences are to be found (see e.g. *HN* 24.133–155). Book twenty-four then turns to 'magical plants' (*HN* 24.156–167) where, again, alphabetical sequences can be observed. Book twenty-five starts with a discussion of the origins of medicine and is devoted to important pharmacological plants (such as moly or *panaces*), with an emphasis on who discovered them: gods, kings, peoples, animals, etc. In book twenty-six, plants are organised according to the diseases they treat best. The diseases discussed are organised in the 'head-to-toe order', starting with skin conditions and diseases of the head, and moving downwards to the feet and diseases that affect the entire body. Book twenty-seven is organised as a herbal, that is, a series of chapters devoted to individual plants. These chapters are organised alphabetically. The philologist Max Wellmann (1889) had observed similarities between this book and Dioscorides' herbal, arguing that both Pliny and Dioscorides had used the works of the pharmacologist Sextius Niger as their source. We would argue that the similarities are in fact more pronounced between Pliny's book twenty-five and Galen's *Simples*, although again more research is needed.

Within the botanical books, Pliny labelled plants 'tree' (*arbor*); 'shrub' (*frutex*); thorn (*spina, spinosa*); herb (*herba*); fruit (*pomum*); vegetable (*holus*); cereal (*frumentum*); legume (*leguminum*), etc. Pliny's attention to classification can be observed in numerous passages, as in the following example: 'Rushes, as they are fragile and grow in marshes, cannot be counted among the bushes (*fruticum*), the thorn-bushes (*veprium*), the stalk plants (*caulium*), the herbs (*herbarum*), or indeed among any type [of plants]' (Pliny, *Natural History* 16.178). The same plant categories can be observed in all Latin works that have an agronomical focus. Both Varro and Isidore of Seville discussed the etymology of these category names, often in a manner that appears fanciful to the modern linguist. For instance Isidore wrote that 'the term "tree" (*arbor*), as well as "small plant" (*herba*), is thought to be modified from the word "field" (*arvum*), because they cling to the earth with fixed roots' (*Etymologies* 17.6.1).

In sum, we would argue that the Roman agronomical writers in general, and Pliny in particular, paid much attention to classification of plants, using various different styles of classification to account for the incredible variety of vegetables and their properties. Much painstaking research remains to be done here, but we hope to have given some solid starting points.

3.4.5 Astrological classifications of plants

Thus far we have encountered classifications of plants according to their parts (morphology), their properties as foods (dietetic), their properties as medicines (pharmacology), or in alphabetical order. Other principles of plant classification were in existence in the ancient world, including some astrological ones. As we saw in Chapter 1, the authors of the astrobotanical treatises argued that the principles of sympathy and antipathy governed the natural world. In their model, each celestial body (sun, moon and the five known planets) and zodiac sign was believed to have particular characteristics (e.g. colour, sex, benevolence or malevolence), and these characteristics were believed to be similar to those of stones, plants and parts of the body. All natural objects sharing important characteristics formed chains – chains of sympathy. Plants linked to a particular celestial body were believed to have healing power over the part of the body on the same chain, especially when that celestial body was in ascendance. However, not all astrologers agreed on what belonged to which chain, hence the variety of our preserved astrobotanical treatises.

The astrobotanical treatise that is best known is *On the Virtues of Herbs*, sometimes attributed to Thessalus or Harpocration (see Chapter 1), whose two books adopt different astrological organising principles. Book one assigns a plant to each of the twelve places of the zodiac: Aries – sage; Taurus – true vervain; Gemini – small vervain, etc. Book two, for its part, assigns a plant to the sun (*hēliotropos*, the 'plant that turns towards the sun', see Amigues 1996c for plants called thus in antiquity), the moon (*aglaophōtis*, the 'shining light', our peony), and each of the five planets.

It is also worth mentioning here a spell preserved in the magical papyri (PGM 3.494–611). Its aim it to establish a relationship with the God Helios, sun god. It is organised by hour of the day. Every hour, the person who pronounces the spell must take the form of an animal, use a certain plant, a stone, and a bird. Here are the first two hours, to give the reader a flavour of this type of document:

> [Spell to establish a relationship with] Helios.... In the first hour you have the form and character of a young monkey; [the tree] you produce is the silver fire; the stone, the *aphanos*; the bird ... your name is Phrouer.
>
> In the second hour you have the form of a unicorn; the tree you produce is the persea; the stone, the pottery stone.
>
> (PGM 3.501–505; translation: W. C. Grese in Betz 1992: 31)

It is easy to dismiss astrobotanical material as fanciful or 'irrational', but in fact the authors of astrobotanical treatises had good knowledge of plants and their powers.

We have now come to the end of our overview of the various principles of plant classification found in ancient botanical texts. Before we conclude this chapter, we need to make a few comments about organisms that are no longer considered to be plants, but that were in antiquity.

3.5 Lichens, fungi and other organisms

In our discussion of ancient principles of plant organisation we have at times mentioned organisms that are no longer classified as plants: lichens and fungi.

These were plants by virtue of their immobility and lack of soul. We shall now give some more detail on the ways the ancients perceived these organisms.[14] Theophrastus' first remarks about fungi relate to their anatomy, and how it differs from that of other plants:

> For not all [plants] have root, stem, branch, twig, leaf, flower, or fruit, or again bark, core, sinews or veins; as in the case of the fungus (*mukēs*) or truffle (*hudnon*). And yet the essence [of a plant] is to be found in these characteristics, and others of the same type.
>
> (Theophrastus, *Enquiry into Plants* 1.1.11)

It is important to note that the reason why modern botanists have removed the fungi from the plant kingdom is not because they lack these characteristics which belong to a plant's essential nature, but because of the chemical nature of their cell walls (which are composed of proteinaceous materials rather than structural carbohydrates) and their inability to produce their own food by photosynthesis due to the lack of the pigment chlorophyll.

In the remainder of the *Enquiry*, Theophrastus mentioned mushrooms on a couple of occasions (*HP* 3.7.6, 4.7.2), but did not refer to the truffle again, perhaps indicating a certain unease in discussing these 'plants' that are so different from his trees, shrubs, undershrubs and herbs. Dioscorides devoted a relatively short paragraph to fungi, indicating that they could be divided into two categories: edible and poisonous (*MM* 4.82). He also briefly mentioned truffles in book two, stating that 'the truffle is a round root (*rhiza*), without leaves, and without a stalk; it is pale yellow; it is dug up in the spring. It is edible and is eaten either raw or boiled' (*MM* 2.145). For Dioscorides, then, the truffle is a root. Galen did not pronounce himself on the status of the truffle (*Simples* 8.20.2, 12.147 Kühn), but noted that fungi are plants (*phuta*) (*Simples* 7.12.25, 12.79 Kühn). In *Foodstuffs* (2.69, 6.655–656 Kühn), the physician considered the truffles and fungi among the vegetables (they are presented between the carrot and the radish). Pliny for his part indicated that fungi can be classified either according to their edibility (safest; safe; poisonous) or according to the type of tree on which they grow (*HN* 22.97). The encyclopaedist also discussed the truffles, which he considered to be marvels (*HN* 19.33).

Theophrastus briefly mentions lichens that grow on the oak (*phaskon*) in a paragraph also listing types of tree galls:

> Only the Turkey oak carries that which some call '*phaskon*' and which resembles rags. It is grey, jagged, and hangs down for a cubit length, like a long shredded piece of linen. It grows from the bark and not from shoots like the acorn.... The sea-bark oak also produces this, but it is blackish and short.
>
> (Theophrastus, *Enquiry into Plants* 3.8.6)

The beginning of this passage clearly refers to *Usnea barbata* L., while the end of the passage may refer to another species such as *Alectoria jubata* Ach. or *Evernia furfuracea* Fries (Amigues 1989: 148).

Theophrastus did not mention lichens that grow on stones. Dioscorides, on the other hand, only mentioned lichens that grow on stones (*MM* 4.53). Galen also discussed stone lichens, stating that: 'the lichen that grows on stones is some sort of moss (*bruon*), and it would be correct to call it a type of plant' (*Simples* 7.11.6, 12.57 Kühn). Finally Pliny noted both stone (*HN* 26.22) and tree lichens (*HN* 26.124). From the point of view of modern botany, lichens comprise a unique growth form, being composed predominantly of a fungal mycelium in which are embedded symbiotic algae. The fungus provides security and shelter, and prevents the alga from becoming desiccated; in turn, the alga produces food by photosynthesis and some of this is used by the fungus. Lichens are not mosses, which are to be classified among plants in the modern system of classification.

Theophrastus also observed growths on plants caused by other organisms, although it is not apparent that he understood that this was the cause of their origin. Thus it is not clear whether he understood the nature of oak galls, although he described them in some detail: oak galls are growths caused by the secretions of the larvae of gall wasps (family: Cynipidae) (see Senn 1942, see also Thanos 2005b). Ancient authors also puzzled over the status of other organisms. They wondered whether coral was a plant or not. For instance, Dioscorides wrote 'it seems that the coral, which some have called *lithodrendron* (literally, the stone tree), is a sea plant and becomes hard when pulled from the bottom of the sea as our ambient air touches it' (*MM* 5.121). Some people in antiquity also believed that amber was the tear of the poplar (see e.g. Isidore, *Etymologies* 17.7.31). We are again brought back to the fluidity of boundaries in ancient conceptions of nature.

3.6 Conclusions

Seneca's *Pumpkinification of the Emperor Claudius* has – perhaps unfortunately – very little to do with pumpkins. Its title is a play on words: '*apocolocynthosis*' (pumpkinification) sounds vaguely similar to 'apotheosis', deification. Claudius, instead of being deified like a good emperor, is transformed into a pumpkin. He joined in this 'honour' other ancient mythical figures who were metamorphosed into plants. These tales of people morphing into plants are only that – stories. However, the ancients really thought the boundaries between categories of living beings were fluid. It was not inconceivable in the ancient world that a plant should morph into another, or for an organism like the sponge to fall between the 'plant' and the 'animal' categories. Philosophers debated throughout antiquity about the very definition of a 'plant'. Their only point of agreement was the lack of locomotion in plants. As a result, they included fungi, truffles, lichens and galls among plants.

Beside difficulties in defining the concept of 'plant', the ancients had to find some organising principles for the immense variety of plants. They had two main options: a classification based on morphology or a classification based on usage. Theophrastus is the key figure with regards to morphological classifications, and traces of his scheme are to be found in all later botanical writings. However, Theophrastus himself acknowledged the limitations of a morphological classification: the category 'herbs' is by far the largest of all and needs sub-divisions. Here, Theophrastus resorted to classification by use, and in particular to the problematic category 'vegetable'. There is no 'vegetable' category in modern botany, 'vegetable' refers to all plants, or is used in vernacular language to refer to edible stuff that is eaten in savoury dishes, as the *Oxford English Dictionary* definition makes clear:

> A plant or fungus cultivated for food; *esp.* an edible part of a herbaceous plant (as a leaf, stem, or root) which typically forms (part of) the savoury course of a meal, served either in a cooked or raw state; such a plant or plant part prepared for eating. Freq. in *pl.*

The pumpkin exemplifies the difficulties in using the 'vegetable' category. Both the Hippocratic text *Regimen* (*Regimen* 2.54, Loeb 4.330 Jones = 6.560 Littré) and Theophrastus classified pumpkins among 'vegetables'. Galen, for his part, in the passage on *Foodstuffs* quoted above, called the pumpkin a 'fruit'. Elsewhere, however, he referred to pumpkins as 'vegetables' or 'vegetable-like' (Galen, *On Foods that produce good and bad Humours* 13, 6.812 Kühn). This is not as contradictory as it appears: in *Foodstuffs* Galen had taken care to define 'fruit' in a dietetic way: the 'dietetic' fruit is something fleshy, edible and usually sweet. In that sense, the pumpkin was a fruit. From an ancient morphological point of view, on the other hand, the pumpkin was to be classified among the herbs, and since the 'herb' category is so large, among the 'vegetables'. In summary, in the ancient world as today, the pumpkin is both a fruit and a vegetable, but it is so for different reasons to our modern ones. Plus ça change...

Notes

1 There is much debate as to the identification of the ancient plant called *kolokuntē*. Some argue that it cannot be a pumpkin (*Cucurbita pepo* L.), which they consider to be a New-World plant. Others suggest that the pumpkin spread from two parts of the world: tropical Asia and North America. See Amigues 2012: 192 for references.

2 Although there appears to have been a more general meaning to *dendron*, one that encompassed other types of plants beyond trees *stricto sensu*, see p. 41 and note 14.

3 On the Aristotelian definition of plants and their relations to other living beings, see Lloyd 1996: 67–82; Sprague 1999.

4 There are some additional points to Aristotle's definition, on which we shall not dwell in this context. What Aristotle meant by 'continuously' is debated by scholars: see Granger 1985.

5 See Lloyd 1966 on the use of analogies in ancient philosophy. For analogies in Aristotle's biology, see Lloyd 1996: 138–159.

6 For a particularly clear explanation of Theophrastus' attitude towards finality, see French 1994: 89–92.

7 See for instance Lloyd 1962; Balme 1962a; Lennox 1980; French 1994: 43–44 and 56–58. For an anthropological approach to the notions of genus and species in botany, see Atran 1987.

8 See André 1954; Deroux 1976; Girard 1988, 1990; Amigues 1999b.

9 Note however that Theophratus calls the citron '*mēlon to Mēdikon ē to Persikon*' (*HP* 4.4.2). See Dalby 2003: 88.

10 It is sometimes argued that the ancients did not carry out experiments. See Lloyd 1979, especially 22–224 for an introduction to the question.

11 On Theophrastus' system of classification, see Strömberg 1937, especially pp. 155–160; Wöhrle 1985: 95–128; Desautels 1988; Amigues 1994, 1998, 1999a.

12 There is very little work on Pliny's systems of biological classification, see Guasparri 2013 for Pliny's zoology.

13 On the Latin word '*frutex*' and cognates, see Ernout 1948.

14 On fungi in the ancient world, see Houghton 1885; Buller 1914–1916; Maggiulli 1977; Sharples and Minter 1983; Grimm-Samuel 1991. On truffles, see Winter 1951.

4

NAMING, DESCRIBING AND DEPICTING PLANTS IN ANTIQUITY

> The cultivated lovage (*ligusticum*): some call it '*panax*' [all-heal]; but Crateuas among the Greeks gives that name to cow-cunila (*cunila bubula*), although others call that plant '*conyza*' [name usually given to fleabane], which is cunilago, while real cunila they call '*thymbra*' [name usually given to savory]. Among us, it has another name, being called '*satureia*' [name usually given savory], classed as a spice.
>
> (Pliny the Elder, *Natural History* 19.165)

4.1 Introduction

There is one major problem facing anyone interested in the study of ancient botany, namely, the identification of plants listed and described in Greek and Latin texts. Identifications and applications of modern Linnaean nomenclature (see pp. xiii and 71–73 for more information) are problematic for various reasons, some of which are illustrated in the text heading this chapter. First, there was much instability and local variability in plant nomenclature in the ancient world. Numerous names could designate the same plant in any particular language (synonymity). There were also problems related to multilingualism in the Hellenistic and Roman worlds. While Greek served as a *lingua franca* of sorts in botanical writing, people did designate plants in their native languages: Latin, Egyptian, Cappadocian, etc. Conversely, the same Latin or Greek name could refer to what we now consider several different species of plants (homonymity). Some plants did not even have names, especially wild ones (anonymity). Finally, and especially in magical contexts, pseudonyms were employed to hide the real identity of a plant (pseudonymity). These then were the four main issues in ancient botanical nomenclature: synonymity, homonymity, anonymity and pseudonymity (for a general introduction to the difficulties in identifying ancient plants, see Reveal 1996). These problems were compounded by the passage of time: Dioscorides' nomenclature differs from that of Theophrastus, for instance (see Riddle 1985: 28). Finally, confusion can ensue from the use of Latin in Linnaean plant names: various ancient plant names – phytonyms – have remained in use as those of modern botanical genera or even families (see Raven 1990; Amigues 2000; on botanical Latin, see Stearn 1973). While Linnaean botanical names sometimes correspond exactly with ancient ones (for instance Greek *murtos*, the myrtle, is known as *Myrtus communis* L. in botanical Latin), in other instances the Linnaean system has borrowed – sometimes quite arbitrarily – ancient names to refer to other, completely unrelated, plants. This is particularly the case for the plants of the New World. For instance, modern *cacti* are not the same as the Greek *kaktos*, a sort of thistle, perhaps the cardoon (see Dawkins 1936: 2).

Second, ancient plant descriptions appear – and we stress the word 'appear' – to lack the systematic aspect of modern ones. They often rely on comparisons with other types of plants rather than on a specialised descriptive vocabulary. There are also instances of plant names without descriptions in Greek and Latin texts, for instance in lists of recipes. Ancient herbal illustrations also seemingly offer little help in identifying plants. Third, some species of cultivated crop plants have evolved and changed since antiquity: the ancient descriptions for these species might not correspond to any contemporary reality.

Despite all these difficulties, scholars since antiquity have attempted to identify plants named in 'older' authorities. Scholiasts and other commentators discussed the identity of plants named in Homer, tragedians, comedians and other poets; physicians such as Galen commented on the herbs named in the recipes of their predecessors, and especially those preserved in the Hippocratic Corpus. The genre of the botanical lexicon (see pp. 102–103) flourished in late antiquity and the Middle Ages. In the early modern period, physicians and other herbalists wrote dissertations on individual plants named in Greek and Roman texts. For instance Georg Wolfgang Wedel (1645–1721), Jena professor of surgery, wrote a *De tetragono Hippocratis* (1688) and a *De holoconitide Hippocratis* (1714). From the beginning of the nineteenth century, the Linnaean system started to make its mark on studies of ancient botanical texts. The English botanist John Stackhouse (1742–1819) published *Illustrationes Theophrasti in usum botanicorum praecipue peregrinantium* (1811), in which each of Theophrastus' plants was placed in one of Linnaeus' sexual categories (polyandra, tetrandra, etc.). Stackhouse then published his *Theophrasti Eresii de Historia plantarum libri decem* in two volumes (1813–1814), where identifications of plants were given, but no Linnaean binomials.[1]

The first to give Linnaean names to plants listed in ancient texts was the botanist Kurt Polycarp Joachim Sprengel (1766–1833), who identified ancient plants in his translation of Theophrastus' works (1822) and in his *Geschichte der Botanik* (1817–1818).[2] His identifications were based, among other things, on the *Flora Graeca* (ten volumes, 1806–1840) compiled by English botanist John Sibthorp (1758–1796; for an introduction to the *Flora*, see Stearn 1976). This was followed shortly afterwards by Heinrich Ludwig Julius Billerbeck's *Flora Classica* (1824), where plants are classified according to Linnaeus' sexual categories and given Linnaean binomials. The late nineteenth and early twentieth centuries witnessed the production of numerous volumes on the flora of the ancients: for instance, the botanist Karl Nikolas Fraas' (1810–1875) *Synopsis plantarum florae classicae* (1870); and Eugen von Halácsy's (1842–1913) *Conspectus Florae Graecae* (1900–1908). Sir William Thiselton-Dyer (1843–1928), fifth director of the Royal Botanic Gardens at Kew, contributed the plant identifications in Liddell and Scott's *Greek English Lexicon*, one the best dictionaries of ancient Greek available (see Liddell *et al.* 1996: vii). Unfortunately, his suggestions are not always reliable (see Raven 1990, reprinted 2000). The French philologist Jacques André studied ancient plant names throughout his career, culminating in the publication of *Les noms des plantes dans la Rome antique* (1985), which examines more than 1,000 ancient Latin names and gives them Linnaean 'translations'. Recently, Suzanne Amigues produced a very thorough edition, translation (into French) and commentary of Theophrastus' *Enquiry into Plants*, which contains the best plant identifications to date (1988a–2006). Finally, Lily Beck, in her recent translation of Dioscorides (2005) gives credible Linnaean identifications.

In addition to these volumes, numerous articles focus on identifying a particular ancient plant, using one or more of the following methods: philological analysis, etymology, botanical observations (in Greece, Italy, Balkans and Asia Minor), and ethnobotany. In particular, we should mention the essays published in the *Actes du colloque international 'Les phytonymes grecs et latins'* (1993) and the numerous studies on individual plants by Jacques André and Suzanne Amigues. Some plants have attracted especial attention, for instance, Homer's moly (see p. 59 of this book), the Cyrenaic silphium (see pp. 54 and 120–121), and the lotos of the lotos-eaters (see Foucaud 1962; Dorie 1967; Herzhoff 1984; Amigues 2004b). The aim of this chapter is not to offer new or revised identifications, but rather to examine the ways in which the ancients named, described and depicted plants. We are looking at these processes from the ancients' point of view, using 'actors' categories', but at times it will be important to compare them with modern ones.

4.2 Naming plants in antiquity

Ancient writers interested in botany were acutely aware of the issues in plant nomenclature outlined above: anonymity, homonymity, synonymity and pseudonymity. They were also interested in the meaning and etymologies of plant names. In this section, after a few preliminary remarks, we will discuss the principles of ancient plant name formation in Greek and Latin, as well as the problems ancient authorities encountered with phytonymy. Our discussion will focus mostly on Greek phytonyms because the Romans often borrowed these, simply transliterating them into the Latin alphabet or slightly adapting them. Jacques André estimated that 30 per cent of plants named in Latin texts had a Greek name (1985: XIII). Greek is a very flexible language, which allows for the aggregation of words to form a new meaning. For instance, the Greek name *pseudodiktamnon* is composed of '*pseudon*' (false) and '*diktamnon*' (dittany) and refers to a plant that is similar to Cretan dittany (*Origanum dictamnus* L.) but smaller (see e.g. Dioscorides, *MM* 3.32). Pliny was aware that Latin was a poor relative of Greek in the field of botany. He noted on various occasions that plants often lacked a name in his writing language, even when in common usage: 'The *rhododendros* has not even found a Latin name among us Romans. They call it *rhododaphne* or *nerium*.... Neither has *rhus* got a Latin name, even though it is of multiple uses' (Pliny, *Natural History* 24.90 and 91). Some of these borrowings, such as *rhus* (sumac, *Rhus coriaria* L.) which appears in the medical writings of Celsus, Scribonius Largus (first century CE, see Scarborough 2008k for references) and others, were well integrated in the Latin language, and had probably been since an early date; others, like *rhododendros* and *rhododaphne* on the other hand, are simple transliterations of Greek words, referring to the plant we know as oleander (*Nerium oleander* L.). They are what linguists call 'xenisms' (see Biville 1993). Now, Pliny expresses a reluctance to deal with 'Greek' topics on various occasions in the *Natural History*, most often in relation to medicine, which he sees as a Greek ill, characterised as it is by complex medicaments and boastful practices (see Nutton 1993; von Staden 1996). However, simple plants, the constituents of Greek compound drugs, remain products of nature, and therefore worth investigating, as expressed in the following passage:

> There are other types [of foliage] which are mentioned only by their Greek names, because ours [i.e. the Romans] for the most part paid no attention to this nomenclature. Even though the majority of these originate in foreign lands, we must examine them, because our discourse is about Nature rather than about Italy.
>
> (Pliny, *Natural History* 21.52)

The naturalist even mentioned a Scythian plant called *anonymus*, which 'found a name by not finding one' (*HN* 27.31). Pliny was very uncomfortable with anonymous plants. This is perhaps because, as classicist Andrea Guasparri argues, 'it is to the names that Pliny links, rather ontologically, the properties of the object being considered' (Guasparri 2013: 350; see also Doody 2011). In Pliny's worldview, an anonymous plant lacks one of its essential characteristics.

While the problem of anonymity was more pronounced in Latin than in Greek, it was not limited to that language. Theophrastus noted that wild plants, which are not commonly used, often remained nameless:

> Most wild plants are without names (*anōnuma*) and few have experience of them. Most cultivated plants, on the other hand, have received names and are more commonly known. I mean plants such as the vine, fig, pomegranate, apple, pear, bay, myrtle, etc. For, since they are in **common use** (*chrēsis*), this leads to their differences being studied.
>
> (Theophrastus, *Enquiry into Plants* 1.14.4, our emphasis)

Naming is taming, and this passage of Theophrastus certainly makes that point clear in relation to plants. Neither Theophrastus nor Dioscorides, nor most ancient botanical authorities, named plants; rather they relied on the information they were given by their informants (see Chapter 2). In this they are very different from modern botanists, who name plants following systematic classificatory principles. Naming always involves a form of classification, but the people who named plants in antiquity, those who used them in their trade or in their daily life, were not interested in universal systematic botanical classifications.[3] When a plant was not useful to them, they simply did not name it.

The names Theophrastus, Dioscorides, Pliny and others gathered from their sources were usually single names, as is the case in modern vernacular languages: e.g. Greek: *murtos*/Latin: *myrtus*, the myrtle. Modern scientific plant names, Linnaean names, on the other hand, are binomials. They are constituted of two words, where the first refers to the genus of the plant, and the second to its species: e.g. *Myrtus communis*, where '*Myrtus*' refers to the genus, and '*communis*' (common) to the species. We saw in Chapter 3 that the ancients sometimes added an epithet to distinguish plants, as in the case of the 'Egyptian bean'. This use of epithets was, however, never systematic. For instance, Galen notes that among the grains, 'some have a simple name, like *sētanion* in Italy ... others have compound names, like the so-called naked barley (*krithē*) in Cappadocia, and the *zeopuron* in Bithynia' (*Foodstuffs* 1.13, 6.520 Kühn). Here we will concentrate on the main name of ancient plants rather than on these epithets.

Ancient phytonyms were not chosen randomly; they usually gave many indications as to the characteristics, properties and origin of a plant. Accordingly, Greek and Roman authors often discussed the meaning of phytonyms and their etymology. In particular, Varro, Pliny, Galen and Isidore of Seville were interested in etymologies. Their explanations give us insight into the principles of ancient botanical nomenclature, although false etymologies are rather common – we will point out a few examples (see Carnoy 1959 for etymologies of Greek plant names). Of course, there were some plants whose names the ancients could not explain from an etymological point of view. They were unable to explain why a vine was called in Greek '*ampelos*', olive tree '*elaia*', or the asphodel '*asphodelos*', even though they sometimes tried (see Amigues 1992; Biraud 1993). Some of these words may predate the arrival of the Indo-Europeans in Greece.

Other plant names had been borrowed from foreign languages. Linguistic studies have unearthed many of these borrowings, some of which are extremely ancient. Thus, ku-mi-no (= *kuminon*, cumin, *Cuminum cyminum* L.) and sa-sa-ma (= *sēsamon*, sesame, *Sesamum indicum* L.), both borrowed from Semitic languages, are attested in Linear B, the syllabic system of writing used for Mycenaean Greek (*c.*1450–1200 BCE) (see Carnoy 1959: 100 and 242; Duhoux 1993: 106 and 110; Casevitz 1993: 404). The name *smurna*/*murrha* (myrrh, *Commiphora Myrrha* Engl.), also Semitic in origin, may have been borrowed during the Mycenaean period too. Indeed Anna Sacconi interprets the ideogram MU (an ideogram is a sign that represents a particular object rather than a syllable), which occurs in Linear B tablets concerned with perfume production, as referring to myrrh (Sacconi 1969; see also Carnoy 1959: 161; Masson 1967: 53–56 for the etymology of *smurna*). Unlike sesame and cumin, myrrh cannot grow in the Mediterranean. If it was available at Knossos, it must have been imported, probably from Somaliland. Other borrowed plants names are only attested in alphabetic Greek (which probably made its apparition in the eighth century BCE), possibly indicating a later introduction, following newly opened trade routes. Thus the word *kassia* (probably referring to cassia, *Cinnamomum cassia* Bl.) first occurs in Sappho (seventh century BCE, fragment 44), and *kinnamōmon* (probably referring to cinnamon, *Cinnamomum* spp., but see pp. 58–60 of this book) first appears in Herodotus (fifth century BCE, e.g. *Histories* 3.107). Both are Semitic words, but *kassia* may have further origins in a Chinese language (Carnoy 1959: 183; Masson 1967: 48–50; Casevitz 1993: 405; Marganne 1996). Foreign phytonyms carried on being adopted into Greek and Latin throughout antiquity. For instance, Celtic phytonyms such as *bricumus* or *blutthagio* appear in medical writings (not only as synonyms), such as those of Marcellus of Bordeaux (fifth century CE, see Stok 2008c for references) (see Polome 1993, see Opsomer-Halleux 1989 more generally for plants names in Latin medical texts up to the tenth century CE).

Most plant names listed in Greek and Latin texts, however, are clearly Greek (or Latin), and the principles behind their application to plants can be discovered by philological analysis. The most common principle of ancient plant nomenclature was to name a plant after one of its characteristics: its morphological aspect, physiological properties, longevity, smell, colour, etc. (see Amigues 1984, 1992; Skoda 1993). For instance, Dioscorides (*MM* 4.88) notes that *aeizōon*, literally, the 'ever-living' (houseleek, *Sempervivum arboreum* L.) was so named in Greek 'because its leaves are evergreen'; Varro (*RR* 1.35) explains that *serpillum* (wild thyme, *Thymus serpyllum* L.) gets its Latin name from the fact that it 'creeps' (*serpit*); and Pliny that the '*cyanus*' (literally 'yellow', blue cornflower, *Centaurea cyanus* L.) is so named because of its colour (*HN* 21.48). While these explanations are accurate, that given by Varro for Latin name of the vetch (*vicia*, *Vicia* spp.), to take one example among many, is rather spurious: 'Vetch is so called from *vincire*, to bind, because it also has tendrils as the vine has, with which, when it creeps up to cling to the stalk of the lupine or some other plant, it usually binds (*vincit*) it … (*RR* 1.31.5)'. Indeed, according to André the origin of the Latin word '*vicia*' is unknown (1985: 271).

When naming a plant after one of its characteristics, analogies with animals, minerals, a body part or any other object, often played a role. Thus the plant named in Greek *thēluphonon* (literally 'female bane', *Doronicum orientale* Hoffm., according to Amigues) was also named *skorpion* (scorpion) allegedly 'because it has a root that looks like a scorpion, and when grated over that animal, it kills it' (Theophrastus, *HP* 9.18.2). Such direct name transfers from animal to plant were, however, quite rare. Plant names were usually derivative, as in the case of the plant named *drakontion* in Greek and *dracontium* in Latin (*Dracunculus vulgaris* Schott), where the suffix *-ion/-ium* indicates that the plant derives its name from the animal *drakōn* (Greek: the serpent). Among the plants named after objects, we can mention that called in Latin *Veneris pecten* (the comb of Venus, shepherd's needle, *Scandix pecten-veneris* L.) after its resemblance to combs (Pliny, *HN* 24.175). As for plants deriving their name from body parts, among many examples, we can list those named *astragalos*, the vertebra (milk vetch, *Astragalus* spp., see Dioscorides, *MM* 4.61) or *orchis*, the testicle (various plants, including the orchid).

A second frequent principle in ancient botanical nomenclature was to name a plant after its habitat. Thus the plants whose names started with the stem *petr-* (from Greek: *petros*/Latin: *petra*, the stone) grew on rocky ground: for instance Greek *petroselinon*/Latin *petroselinum* (parsley, *Petroselinum hortense* Hoffm.) (see Bonet 1993).[4] Other species indicated their liking for water in its various environmental forms. Thus Greek *hudrolapathon*/Latin *hydrolapathum* (water sorrel, *Rumex* spp.) grows near water (*hydr-*, from Greek *hudōr*, water); and Greek *heleioselinon*/Latin *heleoselinum* (marsh celery, wild *Apium graveolens* L.) is a marshland plant (*heleio-*, from Greek *heleios*, of the marsh). The extension of that principle was to name a plant after a region or country. For instance, Dioscorides noted that the plant named in Greek *ligustikon* (lovage, *Ligusticum vulgare* L.) is so named because it grows abundantly in Liguria (*MM* 3.51, see also Pliny, *HN* 19.165 on the plant named *ligusticum*); and Pliny discussed a plant called '*scythice*' which grew around Lake Maeotis, the Sea of Azov around which the Scythians dwelt (*HN* 25.82).

The name '*britannica*', identified with a type of dock (*Rumex* sp.), is more of a mystery. The herb had antiscorbutic properties which were pointed out to the Romans by the Frisians (see Map 3):

> The Frisians, then a trustworthy tribe, in whose territory our military camps were, showed it to us. I wonder what the reason for this plant's name is, unless perhaps, living by the British Ocean, they have named it thus [i.e. *britannica*] just as if it were a neighbour. It was certainly not named because it grew abundantly there [in Britain], for at that time Britannia was independent.
>
> (Pliny, *Natural History* 25.21 (note however that at *HN* 27.2, Pliny says that this plant comes from Britain))

As rightly noted by archaeologist Andrew Fitzpatrick (1991), the Frisians of Pliny's story did not discover the actual plant (which is widely distributed), but rather its antiscorbutic properties. Pliny wonders why a plant was called '*britannica*' before the conquest of Britain, before the time of Germanicus. The name might reflect a Frisian folk tradition: this people may have learnt the properties of the plant from Britain, and hence called the plant after their informants (although Fitzpatrick gives various other possible explanations). Interestingly, in the 1920s, a lead lid inscribed '*ex radice britanica* [*sic*]' was found at the legionary base of Haltern, in Frisian territory. It probably dated to thirty or forty years before the conquest, although its date cannot be ascertained.

Another example of a plant seemingly named after a country is the lavender (*Lavendula stoechas* L.; Greek: *stoichas*/Latin: *stoechas*). Various ancient authors ascertained that it was called thus after the Stoechades Islands, a group of small islands situated some seventy kilometres offshore of Marseilles (ancient Massalia, see Map 3; Dioscorides, *MM* 3.26; Galen, *Antidotes* 1.15, 14.76 Kühn; Isidore, *Etymologies* 17.9.88). Pliny even went as far as to say that the plant only grows in those islands (*HN* 27.131), which is certainly untrue. This is a clear case of false etymology. As noted by Amigues, the ancients would most probably not have named a common plant after some very small, relatively unknown islands of the Western Mediterranean. The name must be linked to the way in which lavender's flowers are disposed, in small, regular vertical lines: the Greek word *stoichos* indeed means 'row, line' (Amigues 1984: 159).

A further principle of ancient nomenclature was to call a plant after one of its properties or functions, and most prominently after its healing properties. One of the best examples is that of Greek *aristolocheia*/Latin: *aristolochia* (birthwort, *Aristolochia* spp.), which derived its name from the fact that it helped women in childbed to deliver their child faster: *aristos* means 'excellent' and *locheia* means 'childbirth', as both Dioscorides and Pliny indicate (Dioscorides, *MM* 3.4; Pliny, *HN* 25.95; on this plant, see Scarborough 2011). Another example would be that of the already-mentioned sumac, Greek *rh(o)us*/Latin *rhus*, which according to Dioscorides, was so named because tanners used it to steep hides (*MM* 1.108.1). The root *rhou-* indeed indicates the red hue of the plant. A further example would be the colourful one found in pseudo-Plutarch's *On Rivers*, which unfortunately cannot be verified: 'There occurs there a herb called "*araxa*", in the dialect of the natives, which translates to "virgin hater". For when the aforementioned plant is found by virgins, it causes a blood discharge, and then it withers' (pseudo-Plutarch, *On Rivers* 23.2). In many cases of pharmacologically active plants, as argued by Alain Touwaide, to mention a name was also to mention its properties (1993). However, it should be noted that antiphrasis, using a word in a sense contrary to its original meaning, and irony are quite common in ancient plant naming. For instance, Pliny writes that '*Holosteon* (all-bone) is a plant with nothing hard about it, the name being an antiphrasis coined by the Greeks, just as they call gall "sweet"'; and that '*Philanthropos* (man lover) is a name which the Greeks in witty sarcasm give to a plant because it sticks to clothes' (*HN* 27.91 and 24.176).

A third principle of ancient phytonymy was to name a plant after the person, real or mythical, who had 'discovered' it or used it; Pliny has a long list of such vegetables in book twenty-five. Thus several plants were called '*panaces*' after Panacea, the daughter of the healing god Asclepius (*HN* 25.30); *mithridateia* was named after Mithradates VI by Crateuas (*HN* 25.62); *gentiana*, the gentian (*Gentiana* spp.) was allegedly called so after a king of the Illyrians, Gentius (*HN*

25.71); *euphorbia* (spurge, *Euphorbia resinifera* Berg.), allegedly discovered by King Juba II, was named after the physician Euphorbus, brother of another famous doctor, Antonius Musa (*HN* 25.77, see p. 25 in this book); and Queen Artemisia (famous queen of Halicarnassus at the beginning of the fifth century BCE) named '*artemisia*' (*Artemisia* spp.) the plant formerly known as *parthenis*, although 'there are some who think that the surname is derived from Artemis Ilithyia, because the plant is specific for the troubles of women' (*HN* 25.73). Again, popular etymologies were common. According to Pliny, the plant *clymenus* was called after the king of that name (*HN* 25.70). While there indeed had been an Arcadian king Clymenus, a plant named after him would probably have been called '*clymen(e)ia*' or '*clymen(e)ion*', with the adjectival suffix -eia or -eion (Amigues 1984: 167). It is more likely that this plant derives its name from the Greek adjective '*klumenos*', famous.

In sum, ancient phytonyms like modern vernacular names, and scientific Linnaean names, evoked the properties, habitats, usages and discoverers of plants. They were full of rich connotations. However, as we noted, in antiquity, plant names varied with time and place: from one ethnic group to the next, from one professional group to the next, and even from one authority to the next. As a result, plants could have numerous appellations, which botanical writers transmitted to their readers. Theophrastus usually separated plants synonyms with the conjunction *ē* (or), for instance, 'the root of *pentaphullon* (literally, "five leaves") or *pentepetous* (literally, "five wings"), for it is called either way' (*HP* 9.13.5). This simple juxtaposition probably indicates that both names, which are very similar in meaning, were in common usage. In other instances, Theophrastus introduced synonyms with the phrase '*ho enioi/tines kalousi*' (which some call…): 'the carob tree (*kerōnia*), which some call the "Egyptian fig"' (*HP* 1.11.2). These phrases may indicate that the second name was less common than the first – although it is important not to over-interpret this phrasing (see Amigues 1988a: xxxviii). Theophrastus learnt some of the synonyms he transmits from the inhabitants of the regions he visited: the Boeotians (see e.g. *HP* 9.13.1), the Macedonians (see e.g. *HP* 3.11.4), the inhabitants of the Ida (of Troad), the Arcadians (see e.g. *HP* 3.9.8), and the Athenians. For instance, he noted that 'some call [this type of ivy] *korumbias*, but the Athenians call it "Acharnian ivy"' (*HP* 3.18.6).

Dioscorides, too, listed plant synonyms, but as Max Wellmann most prominently noted (1898b, see also Popa 2010 for a summary of the scholarship), many other synonyms were added to his plant descriptions in the alphabetical reworking of the *Materia Medica*, maybe as early as the first century CE. Wellmann therefore distinguished between the authentic Dioscoridean synonyms and those of the alphabetical recension – let us start with the former.[5] Like Theophrastus, Dioscorides either separated synonyms with the conjunction *ē* (or), or introduced them with phrases such as '*oi/tines de kalousin*', '*enioi de legousin*' (but some call it), *kaletai* (it is called), etc. (see Touwaide 1993). In general, Dioscorides gave relatively short lists of synonyms. He sometimes explained why a certain synonym was given to a plant, as in the case of the mandrake, which some called '*kirkaia*', after the famous Circe, because of its magical powers: 'The mandrake: but some [call it] *antimimon*, others *bombochulon*, and others *kirkaia* because its root seems to be effective in love potions' (Dioscorides, *Materia Medica* 4.75). Dioscorides inherited some of his synonyms from his predecessors, whom he sometimes acknowledged (see e.g. Crateaus at *MM* 4.74), sometimes not.[6] Other synonyms, he may have collected himself, including several 'ethnic' ones: he lists Cilician (e.g. *MM* 3.68), Syrian (e.g. *MM* 2.167) and Latin synonyms (e.g. *MM* 1.115.5). The Latin synonyms are particularly noteworthy (see Stadler 1898). Dioscorides was writing in a very different world from that in which Theophrastus was writing. Dioscorides' Greek-speaking region (Cilicia) was under Roman rule, and Rome was one of the major trading cities; hence Dioscorides' need to record some Latin plant names. Thus when Dioscorides writes 'the rosemary which the Romans call *rosmarinum* and which wreath-makers use' (*MM* 3.75), he alluded to the Roman taste for garlands and other wreaths, which could retail at high prices.

The synonym lists in the alphabetical recension of Dioscorides, to which we now turn, included many more 'ethnic' names. In total twenty-six nations make an appearance, among which we can mention the Gauls, Sicilians, Dacians, Dardanians, Spanish, Egyptians, Africans, Armenians, Syrians, Marsi and Babylonians (see Bologa 1930; Fortes Fortes 1997; Johnson 2006; Scarborough 2006b; see Map 3 for the location of most of these peoples). Egyptian synonyms are most prominent, perhaps because the compiler of the synonym lists resided there, and/or because Egypt was a crucial player in plant export in the ancient world. However, it is necessary to consider these 'foreign' names with caution. In his study of Etruscan plant synonyms, Kyle Johnson has shown that some 'Etruscan' names are in reality Greek. The synonym lists of the alphabetical recension also include names that were given to plants by the prophets (= Magi), Democritus, Zoroaster and Osthanes, all people whom some botanical authorities considered as 'marginal' (see Chapter 2).

Wellmann suggested that these long lists of synonyms in the alphabetical recension were excerpted from Pamphilus, a first-century CE lexicographer, author of various lexica, and in particular of a work *On Plants* (*Peri Botanōn*, see Galen, *Simples*, preface to book 6, 11.793 Kühn). Pamphilus would certainly fit the bill: according to Galen, this author showed too keen an interest in magic and for listing unnecessary Babylonian and Egyptian plant names. Pamphilus, however, is by no means the only possible contender. Plant lexica were a common genre in the ancient world, and various examples have survived, although they are later in date than the first century CE. Armand Delatte (1939) collected fifteen, to which Margaret Thomson (1955) added a few more – many others could be mentioned and many others remain unedited (see Thomson 1933; Stannard 1971; Touwaide 1999b). More general medical lexica also contained entries giving plant synonyms. For instance, Erotian in his *Collection of Hippocratic Words*, gives some plant synonyms: '*Leirion*: a type of plant, which we call *krinon* [name usually given to the lily]. [Sextius] Niger in his *Materia Medica* says that some call the *leirion, narkisson*' (p. 59 Nachmanson).

Galen too had a strong interest in synonyms, which he listed in his writings on *Simples*, on *Foodstuffs* and in various lexica. Like Dioscorides he borrowed lists of synonyms from predecessors, but he also added much personal information, in particular relating to synonyms from his own region of Cilicia. For instance, he wrote the following about the vetch: 'the name "*bikos*" is entirely usual in our region, and it is only called thus here, but among the Athenians, it is also called *arakos* or *lathuros*' (*Foodstuffs* 1.36, 6.551 Kühn). Earlier in the same treatise, Galen had made some very knowledgeable comments about the name *arakos*:

> We find that the last syllable of the name '*arakos*' is written with a 'kappa' in Aristophanes' *Merchant Ships*.... In our region, they call '*arachos*' the wild plant that is round, hard, and smaller than bitter vetch (*orobos*), and which is found among the fruits of Demeter [i.e. corn]. They pronounce it '*arachos*', not with a 'kappa' in the last syllable, but with a 'chi'.
>
> (Galen, *Properties of Foodstuffs* 1.27, 6.541 Kühn (see Wilkins in Powell 2003: xiii))

In several places, Galen criticised those people who, instead of common plant names, used the old 'Atticising' names, that is, the names found in texts of the classical period, some 600 years before Galen's times (see e.g. *Foodstuffs* 2.9, 6.579 Kühn). In later antiquity, plant lexica were attributed to Galen. Delatte edited a *Lexicon botanicum* (or *Glossarium plantarum*), which is basically a list of synonyms, attributed to the physician, although it is most probably not authentic (Delatte 1939: 385–393).

Most, if not all, botanical and medical writings that are preserved contain synonym lists, either short or long. There is not enough space here to examine synonyms in all these writings – the topic in general deserves further studies – but we shall mention one further example. A couple of Demotic magical papyri (PDM 14.897–910 and PDM 14.966–969), that is, papyri written in a late form of Egyptian, list various ingredients, several of which are vegetable. The papyri give the Greek name of these ingredients, followed by their translation in Demotic (a form of Egyptian script), and in one case a short plant description.

A final problem in ancient botanical nomenclature is that of pseudonyms: false names used to hide the identity of real plants. We are made aware of this issue primarily through one of the magical papyri, which transmits a list of pseudonyms and their 'translation', as well as the following prefatory remarks:

> Interpretations which the temple scribes employed, from the holy writings, in translation. Because of the curiosity of the masses they [i.e., the

> scribes] inscribed the names of the herbs and other things which they employed on the statues of the gods, so that they [i.e, the masses], since they do not take precaution, might not practice magic, [being prevented] by the consequence of their misunderstanding. But we have collected the explanations [of these names] from many copies [of the sacred writings], all of them secret.
>
> (PGM 12.401–407, translation H. D. Betz and J. Scarborough in Betz 1992: 167)

Most plants in the list (many of which are quite common, and not particularly dangerous: dill, wormwood) are named after a body part or bodily fluid, with semen and blood figuring most prominently:

> Tears of Hamadryas baboon: dill juice;
> Blood of Hephaistos: wormwood;
> Hairs of a Hamadryas baboon: dill seed;
> Semen of Hermes: dill;
> A man's bile: turnip sap;
> A pig's tail: leopard's bane;
> An eagle: wild garlic (?);
> A hawk's heart: heart of wormwood.

The choice of these pseudonyms is based on the complex relations of sympathy that some authorities postulated between plants, animals and metals. However, it is difficult to know how often these pseudonyms were used and in what circumstances: magical papyri normally list usual plant names rather than such pseudonyms. At times, these pseudonyms may have led to confusion, as shown by an example studied by historian of ancient magic Patricia Gaillard-Seux (1999). Pliny, in book twenty-nine of the *Natural History* (a book that deals with animal products), discussed basilisk's blood, which was also called *saturni sanguis*, the blood of Saturn (*HN* 29.66). Pliny clearly thought he was dealing with the blood of the famous – mythical – serpent basilisk. However, the name 'blood of Saturn' clearly recalls the pseudonyms given to plants by magicians. Pliny's 'blood of basilisk' may well be a plant.

In sum, nomenclature is indeed an issue when studying ancient botanical texts. However, ancient writers were acutely aware of the problems at stake and put in place some mechanisms to minimise the effects of unstable nomenclature. One such mechanism was the synonyms list; another, to which we now turn, was to describe plants in detail.

4.3 Describing plants in antiquity

Ancient plant descriptions were mostly descriptions of plant parts. In order to provide a good description, it was therefore necessary to define these parts. For

Theophrastus, these definitions were doubly crucial, since as we saw in Chapter 3, he based his classification of plants on their parts. In the first part of this section, we shall determine which parts of plants the ancients distinguished. We will then discuss how ancient authors referred to these parts in their descriptions of individual types of plants.

Theophrastus opened his *Enquiry into Plants* with a detailed list of those parts on which he based his overarching botanical classification. He distinguished 1) the parts which are 'primary (*prōta*), most important, and common to the majority of plants'; 2) the parts from which these primary parts are made; and 3) the annual parts. Let us take these categories separately, starting with the primary parts:

> Root (*rhiza*), stem (*kaulos*), branch (*akremōn*), twig (*klados*); these are the parts in which we may divide [the plant], as if they were members (*melē*), as in the case of animals. For each is distinct (*anhomoion*), and the whole is composed of all these. The root (*rhiza*) is that by which [the plant] draws its nourishment; the stem (*kaulos*) that to which it is led. And when I say 'stem' I mean the part that grows above the earth and which is single. For that is the part that is most common both in annuals and in long-lived plants; and in the case of trees it is called 'trunk' (*stelechos*). By 'branches' (*akremonas*), I mean those parts that are split from the stem and which some call 'boughs' (*ozous*). By 'twig' (*kladon*), I mean the offshoot (*blastēma*) from these [branches] in one occurrence, and in particular annual growth.
>
> (Theophrastus, *Enquiry into Plants* 1.1.9[7])

For Theophrastus, the primary parts are distinct from each other. In Aristotelian terms, these primary parts are an-homoiomerous, that is, they are themselves made of other constituents, such as bark and sap (see p. 106). Theophrastus acknowledged that not all plants have all the primary parts he listed (root, stem, branch, twig), and that his primary parts belonged more particularly to trees. Although Theophrastus was using the Aristotelian distinction between an-homoiomerous and homoiomerous parts, a distinction that originated in the study of animals rather than plants, he stressed the difference between plants and other living beings:

> On the whole, the plant is a thing that is diverse, manifold, and difficult to define in general terms. Take as a proof of this the fact that one cannot find anything that belongs to all [plants], as a mouth or belly [are common to] all animals.
>
> (Theophrastus, *Enquiry into Plants* 1.1.10)

With these provisos in mind, Theophrastus went on to list the components of the primary parts of plants – in Aristotelian terms, the homoiomerous parts:

> And there are other [parts] from which these parts are made, such as bark (*phloios*), wood (*xulon*), and core (*mētra*), for those plants that have a core. And all these are homoiomerous (*homoiomerē*). And there also are parts which are even prior to these, and from which they are made: sap (*hugron*), fibre (*is*), veins (*phleps)*, flesh (*sarx*). For these are principles (*archai*), unless one should call them the powers (*dunameis*) of the elements (*stoicheiōn*), and they are common to all [plants]. Thus the essence and the entire nature [of plants] resides in these.
>
> (Theophrastus, *Enquiry into Plants* 1.2.1)

Next, Theophrastus described the annual parts of the plant:

> But there also are parts which are, as it were, annual, which contribute to fruit generation, such as the leaf (*phullon*), flower (*anthos*), stalk (*mischos*) (this is the part from which the leaf and fruit are joined to the plant). And again there is the catkin (literally, the spongy part, *bruon*) in those plants that have it. And in all cases, the seed (*sperma*) of the fruit (*karpou*). For the fruit (*karpos*) is the sum of the seed (*sperma*) and the pericarp (*perikarpiou*).
>
> (Theophrastus, *Enquiry into Plants* 1.2.1)

Lists are never gratuitous. Theophrastus chose every word in the paragraphs we quoted extremely carefully, so much so that the botanist Edward Lee Greene called them 'the most classic piece of elementary botany in existence' (1909: 62). In these paragraphs, Theophrastus was giving a well-limited, systematic definition to words that had been used in a wider sense, or whose meaning was rather unstable at the time (see Blanc 1993). Let us take a couple of examples, starting with the pericarp. The Greek word *perikarpion* first appears in the writings of Aristotle, who may well have coined it (see e.g. *On the Soul* 2.1, 412b2). It was a useful technical term to distinguish between the seed covering (husk, shell, flesh, pod, chaff, fleshy edible part, etc.), which sometimes as in the case of the citron has several layers (see e.g. Galen, *Foodstuffs* 2.37, 6.618 Kühn), and the seed *stricto sensu*, which had as its main purpose the generation of plants. The *perikarpion* protects the seed and will be discarded in the process of generation. That word, however, does not appear to have 'caught on' outside of Aristotelian circles: Dioscorides only uses the word three times (*MM* 2.110, 3.144, 4.172), and we saw in the previous chapter how Galen called 'fruit' (*karpos*) what Theophrastus would have called *perikarpion*, namely the fleshy, edible part of a fruit. As we will see in Chapter 5, Theophrastus knew that the flower precedes the fruit, and that the flower was somehow linked to the generation of plants. He did not, however, define the flower in any detail. His 'flower' – like that of all ancient authors – is the corolla of modern botany, that is, the colourful petals of a plant. The ancient 'flower' is not the whole flower of modern botany. This explains why some plants that we today classify as flowering were deemed to be flowerless in antiquity. For instance, Theophrastus wrote that:

> There is also a controversy concerning the flowers of some plants, as we have said. For some believe that the oak flowers, as well as the hazel and the sweet chestnut, and again the pine and the Aleppo pine. Others, on the other hand, believe that none of these trees flower, but that catkin (*ioulon*) of the nut trees, the moss-organ (*bruon*) of the oak, and the honeycomb-organ (*kuttaron*) of the pine are all similar and analogous to the fruits of the fig tree that fall prematurely.
>
> (Theophrastus, *Enquiry into Plants* 3.3.8)

All the structures Theophrastus is describing here are in fact what modern botany would call 'flowers'. The catkin of nut trees, for instance, is a cylindrical flower cluster. It may not have petals, but it is a flower organ – nut trees are flowering. Even the fig is a 'false fruit' (pseudocarp), in which flowers and seeds form a single mass. The ancient understanding of a 'flower' was clearly very different from the modern one: the ancient flower was a beautiful appendage, whereas the modern flower is the organ of plant reproduction (see Greene 1909: 83).

While Theophrastus' understanding of the flower was perhaps limited, his reflection on the root of plants was extremely careful. The Greek word '*rhiza*' was particularly polyvalent, designating plant roots, plant bulbs (such as those of the onions or orchids), as well as pharmacologically active plants taken whole. While Theophrastus retained that latter sense in book nine of the *Enquiry into Plants*, which deals with the herbs of the root-cutters, he was careful in distinguishing root and bulb. He wrote that:

> It is not correct to call everything that is under the earth 'root', for if that were the case, the stalk of purse-tassels and chives, and in general any part which is are underground would be a root, and so would be the truffle, the plants that some call *aschion* [perhaps earth-balls, *Scleroderma verrucosum* Bull.], the *ouingon* [perhaps a *Terfezia*], and all other underground plants. In fact, none of these is a root, **for we must define on the basis of natural function and not on the basis of position.**
>
> (Theophrastus, *Enquiry into Plants* 1.6.9, our emphasis)

Some of Theophrastus' followers did use the word '*rhiza*' to designate a bulb, as Galen in his lexicon of Hippocratic words: '*Didumē*: the root (*rhiza*) of the plant *orchis* and a name of the plant itself' (19.93 Kühn).

To Theophrastus, parts should be defined according to their function, and not their position. The function of the root is to draw nourishment. In that sense, it is analogous to the mouth in animals, but as already pointed out (see Chapter 3, and see further Chapter 5), Theophrastus was extremely careful in his use of analogy, avoiding anthropomorphism. That cannot be said of his successors, who stressed equivalence – rather than analogy – between parts of plants and animals. Nicolaus of Damascus, for instance, in *On Plants* summarised Theophrastus' list of plant parts, reversing his order (starting with the homoiomerous parts, followed by the an-homoiomerous), and emphasising correspondences with animal parts:

> Just as in animals there are members consisting of like particles, so also in plants. And every part of a plant is comparable to a member of an animal, so that the bark of a plant is comparable to the skin of an animal, the root of a plant to the mouth of an animal, its knots to the nerves of an animal, and so with the other things it has.
>
> (Nicolaus, *On Plants* 1.3.63, 818a17–23; translation: Droosaart Lulofs and Poortman 1989: 144)

Pliny, for his part, described Theophrastus' an-homoiomerous parts of plants in the following, highly anthropomorphised terms:

> There is a fluid (*umor*) in the body (*corpori*) of trees, which must be understood as their blood (*sanguis*).... And in general, the body of trees, as well as those of other living beings (*animalium*) have skin, blood, flesh, sinews, blood vessels, bones, and marrow. Bark in place (*pro*) of skin ... Next to this most trees have adipose tissues; these are called '*alburnum*' from their colour ... Under this is the flesh, and under this meat, there are the bone, that is, the best part of the timber.... In the flesh of some trees there is pith and blood vessels. It is easy to distinguish between them: the blood vessels are broader and brighter than the pith.
>
> (Pliny, *Natural History* 16.181–184)

Columella for his part described the vine in terms of feet, shoulders, trunk, arms, loins, head and clothes (their fruit and foliage, *vestitae*) (*RR* 4.21.1; *On Trees* 3.1).

At the end of book one of the *Enquiry*, Theophrastus divided plants into groups according to their roots (*rhizai*), nodes (*ozoi*) and leaves (*phulla*), giving examples for each type. For instance, he wrote: 'Some plants have few roots (*oligorrhiza*), as pomegranate and apple' (*HP* 1.6.3); 'some nodes are opposite one another, as those of the wild olive, while others are set at random' (*HP* 1.8.3); and 'some trees are broad-leaved (*platuphulla*), as vine, fig and plane' (*HP* 1.10.4). Theophrastus took full advantage of the flexibility of the Greek language to coin numerous new botanical adjectives such as *oligorrhizos* (that has few roots), *platuphullos* (that has broad leaves), *polurrhizos* (that has many roots), *monorrhizos* (that has a single root), *makrorrhizos* (that has a large root), *bathurrhizos* (that has a deep root), etc. (see Strömberg 1937: 47–94)

When one turns to descriptions of individual types of plants in Theophrastus' *Enquiry*, however, this technical vocabulary is not particularly prevalent. Instead, his descriptions, as all ancient botanical descriptions, very much rely on comparing the parts of one plant with those of another (e.g. the leaves of species x look like those of y) rather than using specific, technical vocabulary (e.g. ovoid). If the reader did not know the other plant, the comparison became meaningless. All this adds up to descriptions that are very difficult to interpret for the modern reader. However, we want to show that, far from being haphazard, they followed rather strict rules and relied on stock comparisons with common plants or objects. Let us consider some examples in Theophrastus' work, as well as in other authors, principally Dioscorides.

At times, Theophrastus concentrated all his knowledge of a plant in one particular chapter, as in the case of the 'Egyptian bean':

> But the [Egyptian] bean grows in the marshes and lakes. The length of its stalk at the longest is four cubits, its thickness is that of **a finger**, it is similar to soft reed without joints. Inside it has partitions which divide it throughout, like honeycombs. On this there is head (*kōdua*), which is similar to a round **wasps' nest**, and in each of cells [*kuttarōn*, the word used is that usually used for the cells of a honeycomb] there is a **bean**, which is slightly raised above it; at most these are thirty in number. The flower is **twice as large as that of poppy**, and the colour is of a deep hue, **like that of a rose**. The head is above the water. Large leaves grow on the side of each bean, equal in size **to a Thessalian felt hat**; they have stalks like the beans.... The root is thicker than **the thickest reed**, and it has partitions like the stalk.... For the root is strong and **not unlike that of reeds**, except that it is thorny.
>
> (Theophrastus, *Enquiry into Plants* 4.8.7–8, our emphasis)

Admittedly, this description relies heavily on comparisons, but with the exception of the mysterious 'Thessalian hats', the objects (honeycomb and wasps' nest), body parts (a man's finger) and plants (bean, reed, poppy and rose) were well known.[8] All these plants occur on a regular basis in ancient plant descriptions.

For other plants, on the other hand, Theophrastus scattered statements throughout his work – the *Enquiry into Plants* is not a herbal, where a single chapter is devoted to each known plant. For example, he first introduced the bramble (*batos*, *Rubus fruticosus* L.) in the definition of a shrub in book one: 'A shrub has many stems and many branches rising from the root; for instance bramble and Christ's thorn' (*HP* 1.3.1). The presence of thorns on the wood is described a few paragraphs later: 'In shrubby and undershrubby plants, and generally in woody plants, one might consider further differences. For the reed is jointed, while the bramble and Christ's thorn are thorny' (*HP* 1.5.3). It is then described as being an evergreen: 'among the shrubby plants, the following [are evergreen]: ivy, bramble, buckthorn, reed, juniper' (*HP* 1.9.4). The spiny leaf-tips next attract his attention: 'Some leaves have slits ... others are somewhat prickly both at the tip and the sides, as those of the kermes oak, oak, smilax, bramble, Christ's thorn and others' (*HP* 1.10.6). It is also noted that the stem itself is spiny (*HP* 1.10.7). For further information on the bramble we then have to go to book three, where Theophrastus makes an ecological remark: 'Christ's thorn grows in rainy and dry places alike, like the bramble' (*HP* 3.18.3). In the next paragraph, Theophrastus enumerates different types of bramble: 'again there are several types, having much variation between them: one grows straight and tall, another bends down and when it touches the earth, it roots again; some call this type *chamaibaton* (ground bramble)' (*HP* 3.18.4). The way in which the berries form clusters is then mentioned: 'it is peculiar to the clusters [of smilax] that they make a row along the sides of the stalk, and that the largest cluster is at the top of the stem, as in the buckthorn and the bramble' (*HP* 3.18.12). In book four, Theophrastus refers to the propensity of the species to grow to some extent in marshes and to be aquatic – perhaps a less-expected characteristic of the plant: 'In some such places [i.e. rivers, marshes, lakes] there are brambles, Christ's thorn and other trees, such as the willow, white poplar, and plane' (*HP* 4.8.1). Theophrastus then adds:

> Even if bramble and Christ's thorn are to some extent plants of the water or living by the water, as they are in some places, their differences from these [i.e. aquatic plants] are nevertheless clear. For we have mentioned both earlier.
>
> (Theophrastus, *Enquiry into Plants* 4.12.4)

Finally, in book six, he returns to the spiny shoot: 'though many trees and shrubs have spines on the shoots, as wild pear, pomegranate, Christ's thorn, bramble, rose, caper' (*HP* 6.1.3). Thus, by the time one has reached the end of the *Enquiry into Plants* one has assembled a quite comprehensive and accurate picture of this species. Assembling such an accumulation of statements is a fairly labour-intensive occupation; however, it was never Theophrastus' aim to be read in this manner. He was not writing a herbal, in which each plant is described separately. To find an example of such a work, one has to turn to Dioscorides' *Materia Medica.*

Dioscorides' actual morphological plant descriptions were usually short, since his main focus was on medical properties and applications (see Touwaide 1999a: 229–234). As historian of medicine John Scarborough notes: 'Although Dioscorides was clearly a very skilled medical botanist, he spends few words on plant morphology, perhaps presuming that anyone interested in such matters would consult Theophrastus' very fine *Enquiry into Plants*' (2010: 7). If a plant was commonly known, Dioscorides often made the following statement 'x is a well-known (*gnōrimon*) plant' (see e.g. *MM* 1.80 for the ash), and left out its description.[9] For instance, he did not describe the garden lettuce because it was such a common vegetable, but he did describe the prickly (wild) lettuce (*Lactuca scariola* L.) because it was not well known (*MM* 2.136). In some instances, Dioscorides described the entire plant rather than its parts: he gave its size (simply stating that it is small or big or giving its height in cubits); its strength (vigorous, sturdy, delicate); its colour; its taste (for smaller plants); its smell; its disposition on the ground (e.g. 'heaped on the ground' *MM* 2.178); whether it had thorns or not, etc. However, in most instances, Dioscorides focused his descriptions on plant parts.

Unlike Theophrastus, he did not define those parts at the start of his *Materia Medica*, neither did he distinguish formally between primary and secondary parts or between permanent and annual ones. The parts that Dioscorides considered are: the leaf (*phullon*, *phullarion*); the root (*rhizon*, *rhizion*, *huporrhizon*); stem/stalk (*kaulon*, *kaulion*); branch (*klados*, *kladion*); shoot (*blastos*, *rhabdos*); twigs (*klōn*, *klēma*); spray (*klōnarion*); axils (*maschalē*); seed (*sperma*, *spermation*); fruit (*karpos*); flower (*anthos*, *anthullion*); thorn (*akantha*); outgrowth (*epiphusis*); umbel (*skiadion*); head (*kephalē*, *kephalion*); and juice (*chulos*). He paid most attention to the leaves and the roots, and least to the flowers (see Touwaide 2007: 42, who argues that leaves served as the basis of some plant classifications in antiquity). Dioscorides' vocabulary relating to branches, shoots and sprays was particularly developed; names were chosen according to the size and/or age of the branch. He also at times used the technical words designed by Theophrastus: many-branched (*poluklados*), narrow-leaved (*leptophullos*), with small fruits (*leptokarpos*), single-stalked (*monokaulos*); without stem (*akaulos*), without fruit (*akarpos*), without leaf (*aphullos*), flowerless (*ananthos*), etc. Again, however, this type of technical vocabulary is not particularly prevalent in Dioscorides' work.

For each of the parts, Dioscorides could describe its size, height, length, thickness, strength, density, shape, consistency, feel to the hand, colour/hue, smell and taste, and disposition. His frequent references to smell and taste are noteworthy: modern botanists avoid sensory comparisons, as they are particularly subjective. Let us consider a couple of examples of Dioscorides' descriptions, starting with that of shepherd's purse (*Capsella bursa-pastoris* L.):

> Shepherd's purse: it is a small herb, narrow with regards to its leaves, approximately a finger in length, bending towards the ground, split at the end, and rather shiny. It sends up a fine stalk, two spans tall, that has few outgrowths, and all around it there is the fruit, rather flat at the top, in which there is a small seed resembling garden cress, quoit-shaped as if it were bruised; it is from that fact that it takes its name. The flower is whitish. It grows on roads, cornices and walls.
>
> (Dioscorides, *Materia Medica* 2.156.1)

Dioscorides only compares shepherd's purse to garden cress (*kardamon*, *Lapidium sativum* L.), which he had introduced in the previous chapter of

Materia Medica (2.155). Otherwise, the description gives actual measurements (e.g. 'two spans tall'), and clearly refers to our shepherd's purse.

The description of liquorice (*glukurrhiza*, *Glucyrrhiza glabra* L.), on the other hand, relies heavily on comparisons:

> Liquorice.... It is a small shrub having two-cubit long shoots, around which there is thick foliage of leaves that resemble **those of mastic**, shiny, and viscous to the touch. The flower is **like that of hyacinth**; the fruit is the size of **the catkin of the plane tree**, rather jagged, having pods **like that of the lentil**, yellowish-red and small. The roots are long, **like boxwood [in colour?], as in the case of gentian**, somewhat astringent, sweet; from them is extracted a juice **like from dyer's buckthorn**.
>
> (Dioscorides, *Materia Medica* 3.5.1)

Dioscorides compared plant leaves to those of mastic (*schinos*, *Pistacia lentiscus* L.) on two further occasions (*MM* 1.96: lime tree; 4.49: stinking tutsan); he used the size of the plane catkins as a measure on two further occasions (*MM* 4.73: head of the thorn apple; 4.136: fruit of the burweed); he referred to lentils on many occasions in his comparisons, and even employed the technical word 'lentiform' (*phakoeidēs*, see e.g. *MM* 4.135); he also alluded to the hyacinth (*huakinthos*, *Scilla bifolia* L.) in his description of the *akanthos* (*MM* 3.17). Clearly, Dioscorides' comparisons of liquorice parts with those of other plants were not haphazard.

All ancient botanical authors followed the same mode of plant descriptions, whether they wrote in the Aristotelian tradition or in the magical tradition; or whether they wrote in prose or in verse. Two further examples will suffice. The first comes from the Hermetic *Cyranides*:

> There are two types of this plant: the first is the white vine [*ampelos*] ... Its twigs, leaves and sarments are **similar to those of the cultivated vine**, but shaggier. It entwines with the neighbouring shrubs, grabbing them with its sarments. It has a **fruit that is grape-like** and yellowish-red, which depilates the skin.
>
> The second is called 'the black vine'.... It has **ivy-like leaves**, rather **similar to those of smilax**, but larger; it attaches itself to trees with its sarments; it **has grape-like fruits**, which are green at first, but become darker as they ripen; it has roots that are black on the outside, but yellowish-red inside.
>
> (*Cyranides* 1.1.106–117, our emphasis)

The compiler of the *Cyranides* compares the parts of these two types of vine to those of extremely well-known plants: the cultivated vine, the ivy and the smilax. Our second example comes from Nicander's *Theriaka*:

> Surely of dark-shaded birthwort one must speak in detail,
> **Ivy-shaped leaves**, like those of the honeysuckle,[10] it carries;
> But its flowers are dyed with crimson, and the smell
> It diffuses is offensive; and the fruit in its midst, **like the wild pear**
> From a cordate pear tree or a common pear, you will choose.
> The root of the female tree is a rounded lump
> But that of the male is long and reaches as much as a cubit in depth
> And in colour it is **like the boxwood** of Oricus.
>
> (Nicander, *Theriaka* 509–516)

This description is certainly more obscure than those found in the works of Theophrastus and Dioscorides – Nicander's works are reputed for their difficulty – but it still relies on comparisons with common plants: ivy, pear and boxwood.

In sum, while ancient plant descriptions relying on comparisons may seem disconcerting to the modern reader, they are by no means random and are actually quite helpful once one gets used to them. As botanist-turned-historian Edward Lee Greene argued, 'this system of ... describing by comparison with types is both natural and not ill adapted to the purposes of phytography. Had it not been so it would not have remained in vogue for two thousand years after Theophrastus' (1909: 104). In addition, ancient herbals were sometimes illustrated: plant depictions, to which we now turn, might have helped in identifying plants.

4.4 Depicting plants in antiquity

The teaching of botany is very dependent on visual aids: it is much easier to describe and explain what something looks like by means of a diagram than by verbal means. Botanical illustrations were also used in antiquity, but were not always seen as beneficial, as Pliny expressed in the following passage:

> Besides these, Greek authors have treated [this topic: the use of plants in medicine], whom we have named in their proper places. Among these, Crateuas, Dionysius and Metrodorus have done so in a manner that is particularly enticing, though one that makes nothing easier to understand but the difficulty of the matter. For they painted images (*effigies*) of the plants and then wrote underneath their effects. But a picture is deceptive when the colours are so many, especially when copying nature, and the various hazards of copying further alter it. In addition, it is not sufficient to paint each plant at a particular period of its life, for it changes its appearance with the fourfold changes of the year.
>
> (Pliny, *Natural History* 25.8)

Pliny is here being particularly ambivalent towards the illustrations contained in the works of the medical authors Crateuas (first century BCE), Dionysius (a very common name), and Metrodorus (first century BCE, see Irby-Massie 2008c for references).[11] Pliny's objections are legitimate: plants do change with the seasons; replicating their colours is no easy task; and copyists do sometimes produce poor quality copies of what may have been a beautiful original plant depiction. Pliny further points to the issues with ancient plant illustrations in his chapter on *dracontium*:

> The plant which the Greeks call '*dracontium*' (dragon plant) has been shown to me in three representations (*effigie*): the first has the leaves of a beet, a thyrsus, and a purple flower; it is similar to the arum. Others have shown to me one with a long root, marked as it were and knotted.... The third description (*demonstratio*) is that of a plant with a leaf larger than that of the cornel, a root like that of the reed, and they say that it has as many nodes as it has years, and that the same goes for its leaves.
>
> (Pliny, *Natural History* 24.150)

The Vienna Dioscorides manuscript contains an illustration of the small *dracontium*. It is closest to the first type of *dracontium* whose image Pliny saw (see Figure 4.1).

John Riddle argued that the original work of Dioscorides was illustrated (1985: chapter 5; see contra Orofino 1991), suggesting that without such illustrations some of the pharmacologist's descriptions make little sense. This is certainly not impossible chronologically, since the authors of illustrated herbals cited by Pliny are all predecessors of Dioscorides; the fashion for illustrating herbals antedates Dioscorides, who had examples to emulate at his disposal.

The most ancient examples of botanical depictions are to be found on fragments of papyrus herbals (see Riddle 1985: 177–179; Fausti 2004; Marganne 2004: 37–42).[12] The earliest is the Tebtunis herbal (see Figure 4.2), named after the Fayum locality of Tebtunis (Ummm el Baragat, see Map 2), of which some twenty fragments are preserved (P.Tebt. 2.679 (descr.) + P.Tebt. Tait 39–41 + P.Tebt.Tait 39 add + PSI inv. 4169 a + b = Trismegistos 63596 = MP^{3}2094).[13] It has been dated to the first or second century CE and was written in columns on a papyrus roll. The illustrations that we have are too fragmentary (stem, leaves, root, flowers) to identify the plants represented, but it seems this herbal was lavishly illustrated. The first editor of the papyrus, J. de M. Johnson, described these illustrations as 'crude and unreal' (1913: 404) – a particularly harsh judgement. The text that is preserved of this herbal relates to the plants chondrilla (Greek: *chondrilē, Chondrilla juncea* L., on which see Dioscorides, *MM* 2.133) and false dittany (Greek: *pseudodiktamnon, Ballota acetabulosa* L., on which see Dioscorides, *MM* 3.32), and bears some similarities to the text of Dioscorides, perhaps indicating the use of common sources. It is interesting to note that numerous medical papyri, both in Greek and in Demotic, were found at Tebtunis, including some Demotic herbals, which are not illustrated (see Hanson 2005).

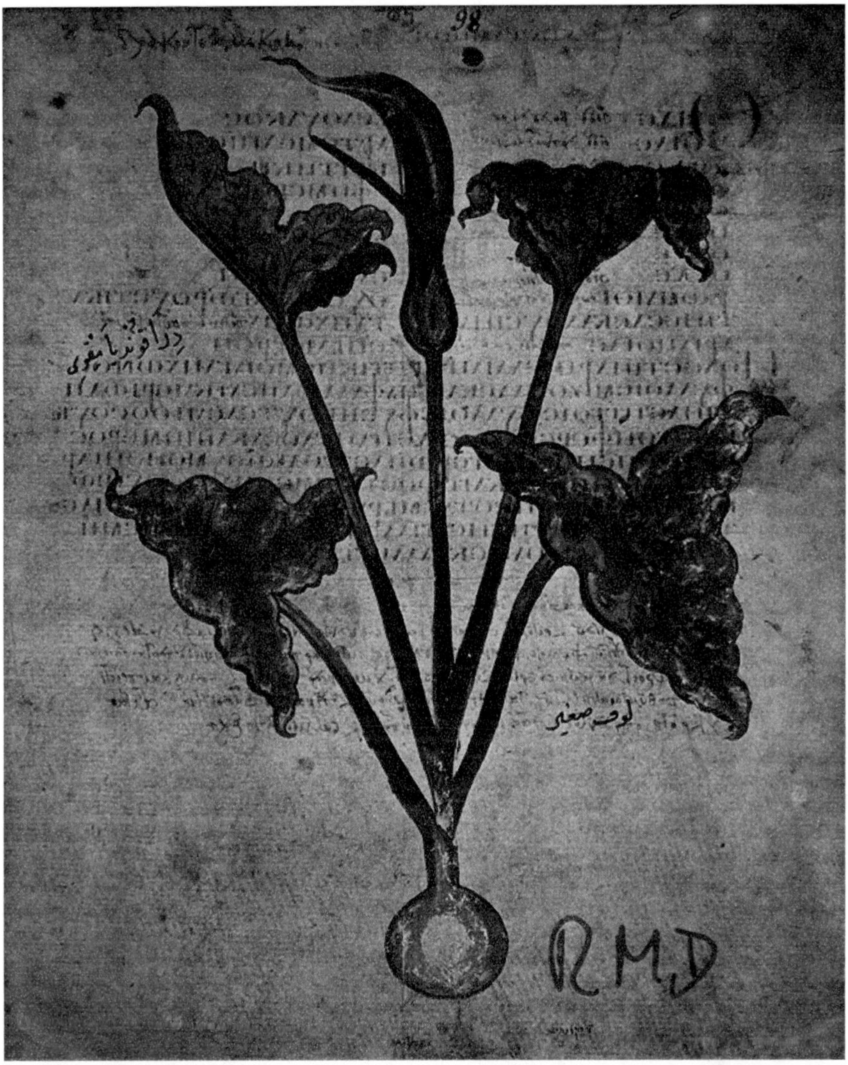

Figure 4.1 The plant *dracontium*, MS Vienna, Österreichische Nationalbibliothek medicus graecus 1 (512 CE), fol. 98r. Courtesy of Österreichische Nationalbibliothek, Vienna.

The second illustrated herbal of which we have fragments is that of Antinoopolis (Sheikh ‘Ibada, see Map 2) (P.Johnson + P.Ant 3.214 = MP^3 2095).[14] Antinoopolis, like Tebtunis, was the site of numerous discoveries of medical papyri. The Antinoopolis herbal consists of the remains of a papyrus codex leaf. Side A (see Figure 4.3) bears the illustration of the plant *sumphuton* (perhaps comfrey,

Figure 4.2 An illustrated herbal on papyrus, second century CE. Papyrus Tebtunis 2.679. Courtesy of the Center for the Tebtunis Papyri, University of California, Berkeley.

Figure 4.3 An illustrated herbal on papyrus (*c.*400 CE) showing the plant *sumphuton* (perhaps *Symphytum officinale* L., comfrey). Papryus Johnson A = MS London, Wellcome 5753. Courtesy of the Wellcome Library, London.

Symphytum officinale L.), together with the name and some information as to how to use the plant in a medicinal context; while Side B (see Figure 4.4) provides the illustration of the plant *phlomos* (perhaps mullein, *Verbascum sinuatum* L.), again with some information as to its medicinal uses.

These papyrus fragments, however, do not allow us to draw significant conclusions on botanical depictions in the ancient world. To do so, one has to turn to late-antique/early-Byzantine manuscripts of Dioscorides and pseudo-Apuleius. These have been studied in detail by Minta Collins, who distinguished two types of plant depictions in such manuscripts:

> Representations of plants fall roughly into two main categories. First are those images in which the primary aim of the artist is to provide a decorative or pleasing representation of a plant, whether singly or in a group, and in which the general appearance of the plant is captured, with or without botanical accuracy. These may be called 'plant portraits'. Second are those images in which the primary intention of the artist is to record or to instruct. These may be called 'plant illustrations'. Obviously the distinctions I make cannot be considered as mutually exclusive, for many plant portraits also record and instruct, and many plant illustrations can be as aesthetically pleasing as the plant portraits. Representations of plants in most manuscript herbals fall into the second category.
>
> (Collins 2000: 27)

The earliest preserved botanical manuscript happens to be one of the most beautiful: the 'Vienna Dioscorides', which we introduced in Chapter 1. It contains a shortened version of the alphabetical Dioscorides, as well as other technical texts. The entire manuscript is lavishly illustrated (in total 383 pictures), although trees, cereals and spices described in Dioscorides' *Materia Medica* were left out from this particular copy, perhaps because they were not considered be decorative enough for this royal gift. In the Vienna Dioscorides, plant illustrations were made before the text was written: the text perfectly frames the images (see Brubaker 2002: 191). Great variations are to be observed in these illustrations: some are naturalistic, others less so. The historian of science Charles Singer suggested that the more realistic illustrations in the Vienna Dioscorides had their origin with the second-century BCE botanical writer Crateaus (mentioned by Pliny in the text quoted above), while the less naturalistic ones were more recent (Singer 1927: 6–17). Indeed, good naturalistic illustrations accompany descriptions of plants where the text of Dioscorides has been supplemented with quotations of Crateuas' text. Riddle and Collins, however, have rejected this hypothesis as too simplistic. There is no evidence that the illustrations accompanying the quotations of Crateuas' text are copies of Crateuas' original, and there are other naturalistic illustrations in the Vienna Dioscorides (Riddle 1985: 190; Collins 2000: 48). The main issue with this hypothesis, however, is

Figure 4.4 An illustrated herbal on papyrus (*c.*400 CE) showing the plant *phlomos* (perhaps *Verbasum sinuatum* L. mullein). Papryus Johnson B = MS London, Wellcome 5753. Courtesy of the Wellcome Library, London.

that it presupposes a stylistic degradation over time: second-century BCE plant illustrators supposedly were more able to draw plants naturalistically than fourth-century CE illustrators. In fact, schematic plant illustrations can at times be more useful to the student of botany than more 'artistic' portraits. In addition, the varying skills of copyists must be taken into account.

Manuscripts containing the *Herbarius* anthology were also sometimes illustrated. The earliest such manuscript is MS Leiden, Bibliotheek der Rijksuniversiteit, Voss. Lat. Q. 9, which may date to the late sixth century CE or to the seventh century CE (see Collins 2000: 167–179). Most other illustrated manuscripts of the anthology fall outside the chronological framework of this book, ranging mainly from the ninth to the thirteenth century CE. In her study of the illustrated manuscripts containing the *Herbarius* anthology, historian of art Heidi Grape-Albers (1977) distinguished three groups: the oldest 'Hellenistic' group comprises naturalistic representations; the second group includes representations that were composed before the end of the second century CE and that are less good than those in the first group, but still clearly recognisable; the third group consists of late-antique representations (third–fourth centuries CE) that are not particularly naturalistic. This model has again been criticised by Collins as being too simplistic; it too implies that late-antique illustrators were less capable than their predecessors of producing naturalistic images (2000: 48–49).

The plant depictions we have considered so far can be said to have a scientific purpose: they illustrate technical botanical texts. Numerous other plant illustrations survive from antiquity. These provide evidence of the cultural, economic and sometimes even political meaning of plants. Thus plants often appear on coins (see Imhoof-Blumer and Keller 1889; Beatty 1974; Baumann 2000). The best-known example is that of silphium on the coins of Cyrene (Libya, North Africa). Silphium was one of the most important exports of Cyrene in the classical period, one that brought much wealth to the Greek colony. Its presence on almost all Cyrenaic coins from the seventh to the first century BCE (see Figure 4.5) indicates how significant that plant was for the economy of the region (the literature on the topic is large, see e.g. Robinson 1927: ccli–cclviii; Gemmil 1973b; Laronde 1996; Koerper and Kolls 1999).

Other plants represented on coins include, among others, a round fruit (melon?) on some coins from Melos (see Map 1); the leaf of celery on coins from Selinus (modern Selinunte) in Sicily (see Map 4), as that city took its name from the wild celery plant (Greek: *selinon*) which occurs in profusion in the area; the rose on coins from Rhodes (see Map 1); the hellebore, which appears on a coin from Pheras; and the pomegranate, which figures on a silver stater from Side in Pamphylia (see Map 2). Scholars have debated whether certain coins from Selge bear representations of styrax, which according to Strabo (*Geography* 12.7.3) grew abundantly in that region of Pisidia (see Map 2; see Amigues 2007).[15] The aim of plant representations on coins is not to offer 'realistic' portraits. However, they may offer some help in identifying plants. For instance, the representations on coins of the mysterious plant silphium can help

Figure 4.5 Silver coin from Cyrene (525–480 BCE), obverse, representing the plant silphium. British Museum, Coins and Medals 1861,1109.1. Courtesy of the British Museum.

identify this plant as an umbellifera, probably of the *Ferula* genus. However, the exact identification of silphium remains one of the biggest puzzles of ancient botany.[16]

Plants also figure on frescoes. One could mention the representation of young girls gathering saffron-crocuses on a Minoan fresco from Thera (Akrotiri, Xeste 3, room 3), whose botanical significance has been studied by Amigues (1988b). More relevant chronologically to us are the landscape and garden Roman frescoes. Some of the frescoes only represent plants as background to mythological or other scenes. Others, however, have plants as their focus. Among these, one can mention the garden fresco in the Casa del Bracciale d'oro at Pompeii (VI.17.42; see Figure 4.6); the garden frescoes decorating room five (a cubiculum) in the Casa del Frutteto at Pompeii (I.9.5), which include paintings of a

Figure 4.6 Mural fresco (first century CE) from the Casa del Bracciale d'oro (House of the Golden Bracelet, VI.17.42), Pompeii, representing trees and birds. De Agostini Collection, Getty images.

cherry tree (south end of east wall) and a lemon tree (north end of east wall); the fresco representing Hercules in the garden of the Hesperides in the caldarium of the Villa Poppaea at Oplontis (Torre Annunziata, near Pompeii); or the magnificent Garden Room at the Villa of Livia at Prima Porta (near Rome; see Figure 4.7), which depicts numerous recognisable trees, including the pomegranate, the palm, laurel, ivy etc. (the Prima Porta frescoes are now preserved at the Palazzo Massimo alle Terme, Rome).[17]

Scholars have shown how these frescoes depict an idealised version of nature: they represent various plants in bloom that do not actually flower simultaneously; they depict plants growing in orderly fashion, with small herbaceous plants in the foreground, fruit trees in the middle-ground, and higher trees in the background. These frescoes are also full of the rich symbolical language of flowers. For instance Barbara A. Kellum stresses how every plant used in triumphal crowns – the crowns worn by triumphant Roman generals – are depicted in the Garden Room at Prima Porta. Particularly prevalent is the laurel, which was especially significant to Livia, the wife of the Emperor Augustus: legend has it that at the time of her betrothal (39/38 BCE) an eagle dropped in her lap a hen holding a sprig of laurel in her beak (Pliny, *HN* 15.136) (Kellum 1994: 218–222). Also remarkable in these garden frescoes are the mixing of wild and cultivated plants, and of plants that did not grow simultaneously in 'real' nature. For instance, as noted by Elizabeth Macaulay-Lewis, the fresco in the Villa of Livia depicts 'the most culturally significant plants of Republican Italy, and of Augustus. Shown blooming simultaneously are plants that actually flowered at

Figure 4.7 Mural fresco (first century CE) from the Villa di Livia at Primaporta, Rome, representing fruit trees. De Agostini Collection, Getty images.

different seasons, in a kind of abundance not seen in the countryside itself' (Macaulay-Lewis 2010: 19). These frescoes, then, do not represent 'true' nature – they represent gardens that are even more beautiful than those which nature, aided by human craft, produces. They may represent individual plants in great naturalistic detail, but they do not depict reality.

One could also mention the numerous representations of plants on the Augustan *Ara Pacis* (an altar to the Goddess Peace, consecrated in 9 BCE) recently studied by art historian Giulia Caneva (2011). She argued that the floral decoration of the *Ara Pacis* carry an important political message rather than being purely 'decorative'. Caneva identified some ninety plants represented on the *Ara Pacis* and uncovered their individual symbolic meaning. As a whole, the message of the *Ara Pacis* is one of metamorphosis and rebirth from roots/tubers lying deep within the earth.

In this section, we have encountered both naturalistic and schematic representations of plants. It is important, however, not to equate 'naturalistic' representations with 'botanically useful' and 'schematic' with 'scientifically useless'. Schematic representations in a herbal dating to the first centuries of the Common

Era may have served a didactic purpose – perhaps a mnemonic one – that is still unclear to the modern scholar. Naturalistic representations on a fresco, on the other hand, may have had some didactic purpose, but they primarily played a symbolic function.

4.5 Conclusions

It is often assumed that ancient botany is somewhat inferior to modern (post-Linnaean botany) in its naming, describing and depicting of plants. This is a rather misguided assumption. Naming of plants was indeed variable in the ancient world, but botanical authors were fully aware of that issue, and put safeguards in place – in particular lists of synonyms – to avoid any possibly fatal error of identification. Ancient descriptions of plants rely heavily on comparisons between plants rather than on specialised technical vocabulary. Modern scholars believe this is a weakness of ancient botany. However, ancient botanical writers were more than able to create such a technical vocabulary, but they preferred not to rely on that vocabulary. Far from being haphazard, their comparisons follow rather strict rules: the number of plants referred to in comparisons is limited. These comparisons would have been easier to understand than technical jargon. Finally, schematic illustrations in herbals are considered to be inferior to naturalistic illustrations. However, as pointed out by Pliny himself, naturalistic plant portraits may look beautiful but they are of little use to botanists, as they only capture one particular moment of a plant's life. 'Simplistic' schematic representations may be much more useful in a didactic context. For ancient botanists were not attempting to replicate nature, but rather to order it, classify it, and bend it to their own rules.

Notes

1 In the introduction to his edition of *Enquiry into Plants*, Hort (1916: xvii) describes this as 'a prettily printed edition with some illustrations; text founded on [the] Aldine [edition]. The editor seems to have been a fair botanist, but an indifferent scholar ... The notes are short and generally of slight value'.
2 See Touwaide 1997 for a bibliography of those who have attempted to identify medicinal plants in Greek texts after the introduction of the Linnaean system.
3 For the expression 'naming is classifying', see e.g. Greene 1909: 122.
4 It should be noted, however, that plants whose name includes the root *petr-* were sometimes named thus because they were believed to dissolve stones in the body.
5 In his edition of *Materia Medica*, Max Wellmann uses a separate section of the *apparatus criticus* to list synonyms which he had found in the alphabetical recension.
6 Dioscorides never mentions Theophrastus as a source for his synonyms, while many of his synonyms are found in Theophrastus' works.
7 For a detailed study of parts of plants in Theophrastus, see Strömberg 1937: 47–136.
8 In his description of the same plant Dioscorides (*MM* 2.106) omits the comparison with hats.
9 The same process is observable in the *Alphabet of Galen*, see e.g. chapter 13 (on *argemone*, the wind rose, p. 154 Everett).

10 Honeysuckle has ovoid leaves, which makes one wonder why its leaves are compared to those of ivy, which have a usual form of three-points. However, some leaves of ivy are more ovoid than others. Still, this comparison is puzzling.

11 For general introductions to the history of botanical illustrations, see e.g. Anderson 1977; Blunt and Stearn 1994.

12 For these illustrated and other herbals on papyri, see also Fausti 1997.

13 For the best edition of this papyrus, see Hanson 2001; see also Johnson 1913; Ryholt 2013. The main fragment is preserved at Berkeley (P.Tebt. 2.679), with further fragments preserved in Oxford. On this herbal, see in addition to the general references given within the main text, Grenfell *et al.* 1907: V and 333; Tait 1977: VII.

14 For the best edition of this papyrus, see Fausti 1998: 43–58. Both fragments preserved at the Wellcome Library, London. On this papyrus, see also Johnson 1913; Singer 1927: 31–33; Leith 2006.

15 According to Suzanne Amigues (on the advice of Pierre Quezel) the coins represent not the tree but the upper part of a press, equipped with bars to manoeuvre it.

16 The bibliography on the identification of silphium is immense. Among the identifications offered, we can mention the following ones: *Ferula tingitana* L. (Sprengel 1807: 39–40); *Thapsia silphium* (no author name provided) (della Cella and Viviani 1819; Cauvet 1875: 13); *Laserpitium gummiferum* Desf. = *Margotia gummifera* (Desf.) Lange (Schneider 1818: 483); *Ferula narthex* Boiss. (Ørsted 1869); *Thapsia garganica* L. (Cauvet 1875: 11); *Lodoicea sechellarum* Labill. (Vercoutre 1913); *Cachrys ferulacea* L. (Manunta 2002). For a summary of the discussions on identification, see Amigues 2004a; Totelin 2014.

17 On these botanical frescoes, see e.g. Gabriel 1955; Francissen 1987; Kellum 1994; Ciarallo 2001; Jashemski *et al.* 2002; Settis 2008.

5

THE LIFE OF A PLANT

5.1 Introduction

For she [the cyclamen] feels (*aisthanetai*) the approach of the river [Nile]. If it should fail,
Through her long roots, just as if she were thinking rationally (*phroneonti logismōi*),
She draws more water, and produces a multitude of fruits.
But never, lying in wait, to gather its fruit,
Will men crave the product of the well-fed cyclamen.
For great is the flood of the Nile and great over the grain
The abundance it spreads.

(*On Egyptian Plants* (P.Oxy. 15.1796), lines 1–7)

Plants are rational beings which consciously control their food intake to produce fruit to feed humans – or at least so they are for the author of this poem preserved on a single papyrus from Oxyrhynchus. This conception of plants is perhaps extreme, even by anthropomorphising and anthropocentric ancient standards, but it nicely introduces the topic of this chapter on the life cycle of plants. In Chapter 3, we saw that the Greeks and Romans considered plants to be living beings, endowed with at least one part of the soul – in Aristotelian terms, with the nutritive part of the soul, which is responsible for generation and growth. It was plain to the senses that several physiological processes in plants were similar to those in animals: plants do generate and grow through the consumption of 'food' and water; they lose leaves as animals shed hair; they grow old, suffer diseases and die. The issue was whether these similarities should be perceived as analogies or as full correspondences. Thus, Theophrastus was far more wary of anthropomorphism in his description of plant physiology than was Aristotle. And Aristotle was a model of restraint compared to Pliny.[1] The analogy between plant and animal physiology worked both ways: the observation of physiological processes in animals could help explain plant physiology; and the observation of physiological processes in plants could help explain animal physiology. For this reason, we will mention a couple of embryological treatises in this chapter, that is, treatises that were written about the human embryo: the Hippocratic treatise *Generation Nature of the Child* and the treatise *To Gaurus* by the Neoplatonist philosopher Porphyry.

In this chapter, we start at the beginning: with the generation of plants. We continue 'chronologically' with plant growth and nutrition; the annual cycle of budding, flowering and fruiting; the ageing process in plants and their death.

5.2 Generation

In the first book of his *On Agriculture*, Varro lists the modes of plant generation – what he calls seeds – as follows:

> Since seeds are in general of four kinds – those given by nature, those which are transplanted from one soil to another as a lively root, those which are torn away from trees and planted into the ground, and those that are grafted from tree to tree – observe what operations you should carry out at each season and in each place.
>
> (Varro, *On Agriculture* 1.39.3)

Varro later divides the seeds given by nature into two categories: visible and invisible, thus hinting at spontaneous generation. Similar lists appear in the works of Aristotle (*GA* 3.11, 761b27–30), Theophrastus (who adds spontaneous generation, see especially *HP* 2.1.1–2), Vergil (*Georgics* 2.9–34), Pliny (*HN* 16.134, 17.58), Columella (*RR* 3.1.1–2, quoting Vergil), Nicolaus of Damascus (*On Plants* 1.6.113, 820b30–34) and Isidore of Seville (*Etymologies* 17.7.73). Theophrastus noted, however, the following: 'Since not all [modes of generation] occur in all plants, it is fitting to distinguish the modes that occur in each type [of plants], and the reasons why this is so, enquiring into the principles of each type's substance' (Theophrastus, *Causes of Plant Phaenomena* 1.1.1). Today, we would only count the first item in Varro's list (seeds given by nature) as plant reproduction. The second and third items in the list are what modern botanists call asexual modes of plant propagation, or cloning, whereby a part (a slip) of a good plant specimen is made to grow secondary roots and is then transplanted. The fourth item in the list, grafting, is the process whereby a shoot (scion) of one plant is attached to the rootstock of another, which has healthy roots growing in the ground. The ancient notion of 'generation', then, differs from the modern notion of 'plant sexual reproduction'. Today, we know that most flowering plants have female (carpels) and male (stamens) sexual organs. Both sets of organs are sometimes present in the same flower, or in two separate flowers of the same individual (monoecious reproduction); while in other cases the female organs are present in the flower of one individual plant, and the male ones in the flower of another (dioecious reproduction). Male plant sexual organs produce pollen, which is carried to the female sexual organs by a variety of means, including bees. As we will see, the ancient understanding of plant sexuality differed completely from the modern understanding, even though authors talked about 'male' and 'female' plants.

In the following sections, we shall review each of Varro's four 'seeds', starting with generation through actual seeds.

5.2.1 Generation by seed

The ancients observed that seeds come in different types (fruits, pips, nuts, grains), shapes (round, oblong, foliated, etc.), and colour (dark or light); and that some plants produce more seeds or more often than others. Theophrastus noted that 'generation from seed (*apo spermatos*) is common to all plants that possess a seed (*sperma*), for all seeds are able to generate', adding in typically Peripatetic fashion that 'if the seed were unable to generate, it would be in vain, since it always has that purpose and is produced for that aim' (*CP* 1.1.1). Nature does nothing in vain. Note that the philosopher talks of 'all plants that possess a seed', thus implying that some plants may not possess a seed. Indeed, in antiquity it was common belief that some plants, and in particular the cypress, willow and poplar did not produce a seed (see e.g. *HP* 3.1.2–3; Vergil, *Georgics* 2.10–13). Aristotle himself had argued that willow and poplar did not produce seed (*GA* 1.18, 726a7). Theophrastus contradicted his master on this particular point, stressing the smallness of these trees' seed. The philosopher suspected that beliefs in the absence of seed originated with farmers who had observed that some plants (and in particular cultivated trees) grow poorly when planted from seed or are never propagated on a farm in that manner (*CP* 1.1.2). However Theophrastus knew that, in theory, that mode of generation was possible, even though it was in practice disadvantageous:

> But since farmers do not use [the seed] for some plants (because these grow to maturity quicker from spontaneous growth, and because for some trees and herbs it is not easy to collect the seeds), some do not quite believe that it is possible for these plants to grow from seed. And yet, as it has been said in the *Enquiry*, in the case of the willow seeds are most plainly evident.
>
> (Theophrastus, *Causes of Plant Phaenomena* 1.1.2)

Elsewhere, Theophrastus played for a moment with the idea that some plants, in particular herbs such as thyme, have no seed. However, he quickly rejected that hypothesis, coming to the conclusions that these plants produce seeds that are 'not evident to the eye, but evident in their power, since the plant is produced by sowing the flowers' (*CP* 1.5.3).

Theophrastus defines the notion of seed in the *Enquiry*, noting that great variety exists, but that in general '[the seed] possesses in itself innate moisture and warmth; and when these are lacking, the seed is sterile (*agona*), in the same way as eggs [are in like case]' (*HP* 1.11.1). In *Causes of Plant Phaenomena*, he adds that strictly speaking the term 'seed' should only be applied to the principle, the starting point, of generation (*archē tou gennan*), and not to the nourishment (*trophē*) and matter (*hulē*) attached to it (*CP* 4.3.4). There too Theophrastus uses the comparison with the animal egg. That comparison was a common one in ancient philosophical treatises, as the question whether plant generation was similar in any way to animal generation was crucial to understanding the hierarchy of nature. It is worth quoting at length a key passage from Aristotle's *Generation of Animals* on this particular issue:

> In all animals that go on foot, the female and male are separate; one animal is female and the other male, even though they are the same in species (*tōi eidei*), just as in the case of humans of both sexes and horses of both sexes. **In plants, however, the [male and female] faculties are mingled together; the female is not separate from the male**; and that is why they generate out of themselves, and produce not semen (*gonēn*) but a foetation (*kuēma*) – what we call their 'seeds'. Empedocles expresses this beautifully in his poem: 'So the great trees lay eggs; the olives first' (DK 31B79). For, just as the egg is a foetation, from part of which the animal is formed, while the remainder serves as food, so from part of the seed the plant is formed, while the remainder serves as food for the shoot and the main root.
>
> (*Generation of Animals* 1.23, 730b33–731a10, our emphasis[2])

In Aristotle's scheme, male and female principles (*archai*) are separate in the higher-ranking animals, those that go on foot (rather than fly, swim or stay immobile). This means that there exist both male and female animals. The male generates out of himself, that is, produces semen; while the female generates in itself, that is, her body shelters and nourishes the growing animal (with some help from the earth, *GA* 2.4, 738b33–35). Since plants cannot move, their generation must be of a different kind, and Aristotle suggests that male and female principles were not separate in plants. What came from this mingling of the male and female faculties – the seed – could not be compared to semen, but was rather like an egg, including both the future plant and its nourishment.

Theophrastus, however, noted that the analogy between plant seed and egg has its limitations, as a seed can take longer to germinate in poorer climatic conditions. He argued that 'animals have their most important principles in themselves, whereas the seed [of the plant] in general gets them [the starting points] from the air' (*CP* 4.11.9, see also 1.7.1). The philosopher thus came to the conclusion that, in the process of plant generation, the earth is by analogy the female element – it is comparable to a womb and can be 'in heat' (*CP* 3.2.6, 4.4.10). Theophrastus' thought is here comparable to the popular ancient notion whereby the earth is female (and females are like the earth).[3] Roman agronomists also stressed the femaleness of the earth. For instance, when discussing the vine, Columella waxed lyrical about the fertility of the earth: 'The fruitful (*alma*) earth each year, as if rejoicing in endless child-bearing (*puerperio*), sends to mortals her breasts (*ubera*), distended with unfermented wine' (*RR* 3.21.3).

Nicolaus of Damascus strongly argued against the idea of male and female principles coexisting in plants, as in his opinion this would imply the superiority of plants over animals:

> We must investigate whether the two sexes are present in one and the same plant, as Empedocles claims they are. As for myself, I do not believe this to be possible, because of a thing that is mingled the simple components must have first existed separately by themselves, and all that is male and female of them is mingled afterwards. The mingling of these takes place only for the sake of generation – accordingly the plant must have existed before the mingling. Moreover, the active and the passive ought not to be together at the same time, and besides, there does not exist any substance in which the males and females are found together in the same thing. If this were so, a plant would be more perfect than an animal, because it would not require anything outside itself for generation.
>
> (Nicolaus, *On Plants* 1.2.39–42, 817a10–22; translation: Drossaart Lulofs and Poortamn 1989: 138 (see also pseudo-Plutarch, *Placita* 5.26, 910c))

Nicolaus is singling out the pre-Socratic philosopher Empedocles as responsible for the theory whereby male and female principles are mixed in plants. He does not mention Aristotle in relation to this theory, perhaps because he wanted to avoid openly criticising the Peripatetic philosopher – in other matters Nicolaus is a follower of Aristotle (see Tortzen 1991: 92). Nicolaus goes on to discuss the comparison between plant seeds and eggs, there praising Empedocles.

In sum, although they disagreed over the exact nature of male and female principles in the process of plant generation, ancient philosophers did not deny their existence. Male and female principles, however, are different from male and female plants, on which we now must say a little. All ancient botanical authors mention 'male' and 'female' plants in their writings; we have encountered several examples already in this book. For instance, some thirty plants and trees are labelled in this way by Theophrastus; sixteen by Dioscorides; and many more by Pliny (see Tortzen 1991: 95). We should of course also mention here the title of the late-antique pseudo-Dioscoridean text *Medicines made from Female Herbs* (*Ex herbis feminis*), even though, as indicated in Chapter 1, this title might be corrupt.

What did Greek and Roman authors mean by 'male' and 'female' plants?[4] Theophrastus introduces the difference (*diaphora*) between 'male' and 'female' most fully in his the first of his books on trees, as this distinction pertains principally to trees, and most particularly to domesticated ones:

> When taking, as has been said, all trees each according to its kind (*genos*), there are many differences. But common to all of them is that by which they distinguish the male and the female, the latter being fruit-bearing, and the former barren in some kinds. In those kinds in which both are fruit bearing, the female has better and more abundant fruits; however some call these the male trees; for some thus call them.
>
> (Theophrastus, *Enquiry into Plants* 3.8.1)

Here, and elsewhere, Theophrastus indicates that the main difference between male and female plants is in their fertility: females are more fertile than males. Because cultivated plants often bear more fruits than wild ones, the former were often considered female while the latter were male (see e.g. *CP* 2.10.1). Some plants presenting characters both male and female were called eunuchs (see e.g. *HP* 4.11.4 on the reed). There were, however, no hard and fast rules in this classification: in some cases, the male plant was more fertile than the female (see e.g. Pliny, *HN* 13.55 on the male sumac tree of Syria); some plants had been known to change sex, as in the case of the seed of female cypresses that produces a male tree (Theophrastus, *HP* 2.2.6); some people argued that both male and female beet came from the same seed (Theophrastus, *HP* 7.4.3); and some authorities, in particular Egyptian ones, reversed the pair, suggesting that the male was always more fruitful than the female (see Pliny, *HN* 16.111; Diodorus, *Historical Library* 1.80.4).[5] The system was clearly not foolproof and there were debates.

While to botanical writers fertility was the most important distinction between 'male' and 'female' plants, to other people working with plants, other differences were more significant. Thus, Theophrastus tells us that for woodcutters, wood quality was the distinguishing factor between male and female:

> This *diaphora* between male and female, as the woodcutters (*hulotomoi*) say, is more or less common to all trees. For every male tree, when it is hewed, is shorter and more twisted, harder to work and darker in colour; the female tree, for its part, is taller.
>
> (Theophrastus, *Enquiry into Plants* 3.9.3)

From other passages, we gather that female trees were generally more amenable to woodwork than male ones (see e.g. *HP* 1.8.2, 5.4.1, see also Pliny, *HN* 16.47). In ancient thought, female animals (and in particular women) were seen as softer than hard men.[6] This, however, was seen as negative: women were softer as a result of their indoor life devoted to gentle occupations such as wool-work, in contrast to the hard outdoor work of men. In the case of trees, on the other hand, female softness appears as a positive – or at least it does to those who fell those trees. Ancient historian Lin Foxhall has attempted to explain this puzzling fact. She noted that ancient woodmen worked with wild trees, trees of the forest. The ancients saw wilderness in general, and forest wilderness in particular, in negative terms: the forest is a place of danger. Woodmen were of an inferior social status in the ancient world. In ancient literature, they are sometimes compared to trees of the wild forest (see in particular Aristophanes, *Acharnians* 178–185). According to Foxhall, woodmen, and the male trees they felled, were associated with a 'sub-standard variety' of masculinity, but any masculinity was superior to feminity. She concludes that:

> Much of the male/female classification of trees is reported or contextualised as central to the usage of woodmen and foresters. The paradigm is constructed in terms of censure and inferiority in the setting of a socio-economic group which was itself marginal and low-status. It is possible that such groups as charcoal burners, lime burners, woodcutters, resin tappers, pastoralists [people who looked after livestock in the forest] and the like themselves internalized and redefined such masculine values as positive, because such groups saw themselves as marginal in several senses to the polis.
>
> (Foxhall 1998: 68)

The division of plants into male and female, then, reflected ancient conceptions of gender, if not always the most 'orthodox' view of gender. Male and female plants, however, did not unite sexually. Since plants are immobile, the ancients did not believe they engaged in sexual intercourse. There was, however, one significant exception, that of the palm, in which contact between 'male' and 'female' led to fertilisation. Mesopotamians had practised what we call 'pollination' of the date palm for centuries before the Greeks became aware of the process (the first Greek to note it in writing was the historian Herodotus, see pp. 53–54). Farmers in Mesopotamia gathered the spathes of the 'male' tree and shook them over the 'female' tree (see Porter 1993).

Theophrastus explained this process quite well, stating that the male palm 'helped' the female to ripen its fruits. He called the phenomenon a mingling (*mixis*) of the sexes, a sexual union (*HP* 2.8.4). Later authors described the process much more vividly – and in extremely anthropomorphised terms. Pliny the Elder's description is filled with sexual allusions:

> For the rest, they say that, in a palm-grove of natural growth, the female trees cannot give birth (*gignere*) without the males, and that many females bend (*pronas*) towards a single male and sway to and fro (*nutare*) with their caressing (*blandioribus*) foliage [hair], and that the male, with his prickly (*hispidum*), erected (*erectis*) foliage, weds (*maritare*) the rest of them with his exhalation, his very appearance, and also with his dust (*pulvere*). And when the male tree is felled, the women become barren (*sterilescere*) in their widowhood (*viduvio*). And the understanding of this sexual union (*veneris*) is such that man has devised a way to bring them together (*coitus*) by means of the flower

> and down collected from the males, and even sometimes by merely sprinkling the females with their dust/pollen (*pulvere*).
>
> (Pliny, *Natural History* 13.34–35)

While Pliny's version is quite erotic, the version found in the *Geoponika*, extracted from Florentinus, a third-century author (see Rodgers 2008b for references), focuses more on longing than lust:

> One palm loves the other, and loves it bitterly. Indeed the female loves the male, as Florentinus says in his *Georgics*, and there is no end to her desire (*pothou*) until she is in the presence of her beloved (*erōmenos*). For the tree appears to stoop, and not to carry itself, and ceases to bear fruit. It does not escape the farmer's attention that she is in love, but he does not know with which tree. For that reason, he touches many palms and then returns to the lover (*erōsan*), and touching her with his hand, she appears to respond as if to a kiss (*philēmati*). She indicates the tree that is the object of her desire by nodding her spathes, or as some would say her hands, for she looks towards him and turns herself towards him from her roots upwards, as if in eagerness. And a remedy (*akos*) for her love is found when the farmer touches the male and brings his arms close to the lover, and especially if he brings the flower from the male spathe, and inserts (*entēhsei*) it into the head of the lover. For this soothes her love, and for the rest, the palm becomes resplendent and bears the most beautiful fruits.
>
> (*Geoponika* 10.4)

The love of the palms became a literary trope (see Pease 1927). A shorter version of the story is also to be found in the second-century Greek pastoral novel by Achilles Tatius, *Clitophon and Leucippe* (1.17), and in the work of the fourth-century, Latin-writing historian Ammianus Marcellinus (24.3.12). The story therefore appears to have circulated widely both in Greek and Latin.

From the point of view of modern botany, date palms are dioecious: male flowers and seed-producing female flowers are on separate trees and need to be brought together for fertilisation to happen. The ancients had clearly come to an empirical understanding of the dioecious character of the palm, and described the process in sexualised terms. However, it must be stressed that they only talked about the generation of the palm in such terms, and never developed a full theory of pollination. The love of the palms is somehow the exception that confirms the rule.[7] We will see later that a form of sexual intercourse was believed to take place in the process of grafting (see also below for comparisons between fertilisation of the palm and caprification).

5.2.2 Generation from root, stem, branch or slip

The second and third 'seeds' in Varro's list (from root and from pieces torn off) are both what modern botanists would call asexual modes of propagation, that is, modes whereby a part of a plant grows secondary roots and is then replanted independently. The Romans sometimes used the word *propagatio* to refer to their way of 'reproducing' trees (see e.g, Cato, *RR* 51; Columella, *On Trees* 7.6), albeit not frequently. There were – and still are – many modes of asexual plant propagation: by means of suckers, branches, roots, leaves, or even pieces of wood (see e.g. the list in Theophrastus, *HP* 2.1.1). Suckers etc. can either be torn away from the parent-plant to be made to grow roots, or in some cases they can be bent and made to grow roots while still attached to the parent-plant. This latter technique is called 'layering'. Aristotle found it interesting that parts torn away from plants could survive (as this does not happen in the case of the vast majority of animals) and generate. He deduced from that fact that plant 'seed' is produced by the entire plant rather than by any particular part (see e.g. *On the Soul* 1.4, 409a9–10). More practically, Theophrastus explained that the parts of plants used for propagation must not be too dry, 'for the [part] destined to sprout independently must always possess the moisture necessary to life, and it must also be able to retain it' (*CP* 1.1.3).

The ancients considered propagation methods to be 'less natural' modes of generation than generation from seed – modes that were matters of 'technique or choice' (Theophrastus, *HP* 2.1.1). Indeed these modes were used in particular to reproduce economically important fruit-bearing trees such as the vine, the apple, pear, pomegranate, fig, and trees grown on estates for their wood. Pliny, while also observing that these methods applied to cultivated plants, was eager to stress that they had been taught by Nature. Thus, he argued that:

> Nature has also shown [how to make] nursery gardens, as from the roots of many trees, there sprouts a cluster of shoots, which are destined to be killed by the mother that begets them: for her shadow oppresses her disorderly multitude.
>
> (Pliny, *Natural History* 17.65)

Pliny goes on to explain, in anthropomorphised terms, that a nurse (*nutrix*) must be found for these sprouts in the nursery (*seminarium*) where they will be able to mature (*adolescere*) (*HN* 17.66). He also stresses how nature has taught the art of layering, which is most visible in the case of the bramble, whose shoots plant themselves in the ground without human help (*HN* 17.96).

Since growing from slips or layers were important methods from an economic point of view, the Roman agronomists devoted much attention to them, starting with Cato. They described in great detail how to choose the best slips, using anthropomorphised vocabulary: the plant from which a slip is taken is the mother (*mater*); the slips are her offspring (*partus*). For instance Columella writes that:

> If from these [vines], as we said above, we select not those parts which are capable of generation (*genitalia conceptu*) and pregnancy with young (*fetu gravida*), but rather their coverings and parasols, which are deprived of fruit, we will toil, so to say, for shade and not for the grape harvest.
>
> (Columella, *On Agriculture* 3.10.12)

Columella also indicated that the slip should be young, but not in infancy (*aetas puerilis*: *RR* 3.10.16). In one passage, he even compared the selection of slips in the cultivation of Aminean vine to the rearing of race-horses (Columella, *RR* 3.9.4–5). Finally, Columella gave the Greek name *spadōn* (eunuch) to shoots that have never borne fruit and will therefore not be suitable for propagation (see e.g. *RR* 3.10.15).

Cato described in detail the layering of trees (*RR* 51–52, 133), which sometimes required pots or baskets with holes at the bottom that were buried in the ground (see also Pliny, *HN* 17.97–98). A plant that is grown in a buried vessel requires less water than one that is planted directly in the soil. After a couple of years, the basket was cut or the pot broken to allow the plant to develop fully. Archaeologists have found pots with holes that were most probably used for the purpose of layering (the technical name for these pots is '*ollae perforatae*'). They were first discovered on the site of the Hephaisteion at Athens in the 1930s. Since then, examples were found, among other places, at Pompeii (in the House of the ship Europa I.15.3), Petra, and London (the pots found in London were produced in Eccles, Kent; see Map 3). The known instances range from the first century BCE to the second century CE.[8]

Cato also described the setting up of nurseries, where the slips were made to grow to a reasonable size and strength before being transplanted.[9] He indicated that nurseries must be located in the most fertile part of the farm (*RR* 46).[10] He also explained how to transplant the cutting once it has grown to the right size, using boxes, pots or baskets with holes at the bottom (*RR* 28). Again, these pots were sometimes buried in the soil with the plant that they helped transplant.

All other Roman agronomists included information on layering, nurseries and transplantations in their works, although they at times disagreed with Cato. Thus Columella warned against Cato's advice to place a nursery in highly fertile soil (*RR* 3.3.2; see also Theophrastus, *CP* 3.5.2). If one uses the most fertile part of a farm for the nursery, once transplanted into poorer soil, the young plant will die or find it difficult to adapt. The good farmer transplants from poorer ground to richer ground. Palladius too noted that 'trees like people should be transferred from bad places to better ones' (*OA* 1.6). We will discuss transplantation of plants from distant regions in Chapter 6.

5.2.3 Grafting, epiphytes and conjoined trees

The fourth, and final 'seed' in Varro's list is grafting (see Figure 5.1), which involves inserting part of a plant (the scion) into another one that is growing

Figure 5.1 Grafting of trees on a Gallo-Roman calendar mosaic (third century CE) from St-Romain-en-Gal, France. Saint-Germain-en-Laye, musée d'Archéologie nationale et Domaine national de Saint-Germain-en-Laye 12–553050. Courtesy of RMN-Grand Palais (musée d'Archéologie nationale), photo by Michel Urtado.

roots in the soil (the rootstock).[11] It was – and still is – used in the propagation of economically important plants. The Romans were fascinated (obsessed even) by the process, devoting large sections of their agronomical writings to the topic. Palladius even wrote a short elegiac poem (170 verses), *On Grafting* (*De insitione*), on the topic. The ancients practised grafting because 'an engrafted (*insita*) tree is more fruitful than one that is not engrafted, that, is, one which is planted in the form of a branch or a small plant' (Columella, *RR* 5.10.6, see also Theophrastus, *CP* 1.6.10). Grafting techniques also led to the creation of new types of fruits, as explained by Pliny: 'The rest of the fruits of trees can hardly be listed by their appearance or form, even less by their flavours and juices, so often have they been intermingled (*permixtis*) and grafted *(insitis*)' (*HN* 15.35).

In grafting, one needed, however, to pay attention to the quality of the rootstock. For instance, to produce nicely flavoured pears, it was best to graft a pear shoot on a cultivated pear than on a wild one (Varro, *RR* 1.40.5). In general, grafting on wild plants was not recommended; however grafting of the cultivated olive on the oleaster was commonly practised in ancient Greece (Amouretti and Brun 1993: 128; Foxhall 2007: 109–111). Authors described several methods of grafting: inserting the graft between the bark and the wood; inserting the graft into the pith; opening a cleft through the middle of the trunk, keeping it open with a wedge; etc. (see e.g. Columella, *On Trees* 26; Palladius, *OA* 3.17). Columella even claimed to have invented another grafting method by means of branches (*RR* 5.11.8–11). Ancient authors also discussed bud-grafting, a grafting method where a bud is used instead of a scion. In fact, most authors only distinguish two types of grafting: grafting *stricto sensu* (Greek: *emphuteia*; Latin: *insitio*; literally: insertion) and bud-grafting (Greek: *enophthalmismos*; Latin: *inoculatio*; literally: inoculation). Theophrastus describes these two processes as types of sexual unions:

> Trees then grow and come into being in the way just mentioned. For grafting (*emphuteiai*) and bud-grafting (*enophthalmismoi*) are, as it were, types of [sexual] union (*mixeis*), or at all events, they are generations (*geneseis*) of another type, which must be dealt with later.
>
> (Theophrastus, *Enquiry into Plants* 2.1.4)

In *Causes of Plant Phaenomena*, he further explains the sexual nature of grafting: in the case of budding, a generative liquid helps the process:

> It remains to talk about the modes of generation [in other trees], namely grafting (*emphuteias*) and budding (*enophthalmismous*).... For the graft (*emphuteuomena*) uses the [tree] like earth. And budding too is a type of planting (*phuteia*), not simply a juxtaposition (*parataxis*); here however it is clear that what produces the sprout and the fruit is the generative (*gonimos*) liquid (*hugrotēs*) whose presence allows the eye to join to the other [plant], and thus having its food, it produces its own sprout.
>
> (Theophrastus, *Causes of Plant Phaenomena* 1.6.1)

Pliny too compared grafting to a sexual union, where scion and rootstock must eagerly unite (*HN* 17.134). In the opening lines of book sixteen of his encyclopaedia, Pliny referred to grafting and budding as 'adoption' and 'marriage' respectively (*HN* 16.1). Palladius also used the marriage metaphor from the very beginning of *On Grafting*, and presented some of his grafts in highly sexualised terms. Finally, Porphyry in his embryological treatise *To Gaurus* used grafting vocabulary to describe human generation – animal generation is closer to grafting than it is to generation by seed (paragraph 10.1–2, see Wilberding 2008). Porphyry compared both embryo and womb to plants: the embryo is like a scion, and the womb is like a rootstock.

Not all ancient descriptions of grafting are sexualised, however. For instance, the author of the Hippocratic treatise *Nature of the Child* gave the following explanation of the grafting process, which is subtly different from that of Theophrastus:

> 'Eyes' that are inserted into trees from other trees, and that grow into trees within the trees in which they live, bear fruits that are not similar to those of the tree onto which they are grafted. This happens in the following way. The first thing that happens to the 'eye' is sprouting, for at first it takes its nourishment from the tree from which it was removed, and then from the one in which it has been inserted. When it sprouts in this way, **it sends fine roots from itself into the tree** [onto which it is grafted]. First it gets [nourishment] from the moisture (*ikmados*) contained in the tree onto which it is grafted. But then after a time, it sends out roots into the earth through the tree into which it is inserted, and it obtains moisture (*ikmada*) drawn [directly] from the earth, and this is how it gets its nourishment. Thus it is no wonder that grafts carry different trees from the [host] trees, since in reality they live from the earth.
>
> (Hippocratic Corpus, *Nature of the Child* 15, Loeb 10.76–78 Potter = 7.528 Littré, see Lonie 1969, our emphasis)

Unlike Theophrastus, the Hippocratic author tells us that the scion grows roots into the rootstock, through which it directly accesses the earth. Now, this is not botanically correct: in a true graft, a scion does not grow roots, but rather the tissues of the scion and rootstock join together. Either the Hippocratic author is mistaken (which is quite possible – his botanical knowledge is not particularly good, whatever standards we apply), or he is talking about 'roots' in a non-literal sense, or he is in fact referring to an entirely different process – an epiphytic process. Indeed, in the case of an epiphyte, that is, a plant that grows onto another, roots do grow. While modern botany clearly distinguishes between grafting and growing epiphytes, ancient botany did not always do so. This is perhaps one of the reasons why naturalists did not agree whether all grafts are possible or not. Theophrastus (*CP* 1.6.2) and Nicolaus (*On Plants* 1.6.114–115, 820b34–821a2), for instance, correctly assert that a scion must be similar in nature to the rootstock for the graft to work. In modern terms, grafts usually work best between individuals of the same species; they sometimes work between members of the same genus or between members of the same family when their genera are closely related. Closely related genera blend together in the case grafts between apples (*Malus* spp., family: Rosaceae; subfamily: Amygdaloideae; tribe: Maleae; subtribe: Malinae) and pears (*Pyrus* spp.; family: Rosaceae, subfamily: Amygdaloideae; tribe: Maleae; subtribe: Malinae), as presented in the following passage of Varro:

> For instance, an oak cannot take a pear, even though an apple can take a pear.[12] This is important to many people who pay much attention to the haruspices; for these say that as many types (*genera*) are grafted (*insita*) into a single tree will produce as many bolts, when the tree attracts lightning, even though the stroke was a single one.
>
> (Varro, *On Agriculture* 1.40.5)

This passage is by the admission of the Loeb translators 'quite unintelligible, though much ingenuity has been spent on it' (Hooper and Ash 1934: 268–269). In the first half of the sentence, Varro asserts that not all grafts are possible, and then goes on to say that some people believe that grafting several types of plants onto one individual may bring bad luck (see Lowe 2010 for the symbolic value of grafting). Whether that second half of the statement contradicts the first, namely implies that an oak can take pears, is rather unclear. Theophrastus, too, notes that phaenomena such as a terebinth growing on an olive, the so-called octopus-plant growing on certain trees, or a bay growing on a plane tree, were interpreted by some as rarities and portents (*CP* 2.17.4). Theophrastus' own interpretation, however, is non-superstitious: a seed has fallen on a spot 'that has become earth-like through decomposition' (see also *CP* 5.4.5). In other words, Theophrastus knows that these phaenomena are not grafting phaenomena. He also notes that growing a plant in another can produce excellent results, as in the case of rue being grown in fig trees: 'and it comes to pass that the juice [of the fig], in addition to feeding [the rue], on account of its heat, provides some help, when needed, just as when ashes are scattered [around rue]' (*CP* 5.6.10).

Unlike Theophrastus, Nicolaus and Varro, Columella maintains that all grafts are possible:

> But since the ancients denied that any kind (*genus*) of slip can be inserted (*inseri*) in any tree, and established as a law that only those slips can unite that are similar in bark, rind and fruit to those trees in which they are inserted, we have thought it useful to destroy this false opinion, and to hand down to posterity this method whereby any type of slip can be inserted into any type of tree.
>
> (Columella, *On Agriculture* 5.11.12 (see also *On Trees* 27))

Columella then explains at length how to engraft figs and olives. From the point of view of modern biology, the fig (*Ficus* spp., family: Moraceae) and the olive (*Olea europaea* L., family: Oleaceae) are completely different trees – they are not even from the same family. It is extremely unlikely that a graft between these two plants could have worked. Columella must be describing some other type of process. Perhaps the cue is in one of the sentences of Columella: 'after an interval of three years, the fig will coalesce (*colescet*) with the olive'. Three years (or even two if counting inclusively) sound like a long time for a graft to take. Now, Theophrastus also describes a process of coalescence in trees (*sumphusis*). This process is different from the *mixis* that happens in grafting:

> But each [shoot] transmits its food separately so that, this not being a mixing up (*epimignumenēs*), each produces its own fruit, which is what rivers do when they meet each other, like the Cephisus and Melas in Boeotia: each flows in its own current ... And the same, or something similar, happens when the same tree bears manifold pomegranates or apples. For they [farmers] soften (as it were) the shoots with the mallet, so that they may coalesce (*sumphuōsin*) because of the bruising; then they bind them together and plant them. Then, in one way, the tree that results from this coalescence (*sumphusei*) is a unity, but each part preserves its own type, drawing and concocting its own food, and sharing nothing but the coalescence (*sumphuseōs*).
>
> (Theophrastus, *Causes of Plant Phaenomena* 5.5.2–3)

What Theophrastus is describing appears to be the conjoining of trees.[13] That process can indeed occur between trees of different genera and looks very similar in its results to grafting. However, from the point of view of modern biology, the processes are different.

In sum, the ancient notion of grafting differed strongly from the modern one, in particular to include what we would call epiphytes (in which according to modern botany, the plant growing on another are not biologically linked) and conjoined trees. We could multiply the examples of ancient grafts that are not considered as such by modern biology, but we leave this topic with a passage in Plutarch's *Table Talk*, where the historian and philosopher presents a dialogue between Soclarus, Crato and Philo on the topic of 'grafting':

> As Soclarus was entertaining us in his gardens, which are surrounded by the river Cephisus, he showed us trees extremely varied by the grafts they took. For we saw olives springing up from mastic trees and mulberries on myrtles. There were also oak trees carrying good pears, plane trees laden with apples, and fig trees grafted on mulberries. And there were other mixtures of trees that were strong enough to carry fruits. Some playfully noted that Soclarus was nourishing kinds of creatures that were stranger than the sphinxes and chimaeras of the poets.
>
> (Plutarch, *Table Talk* 2.6.1, 640b–c)

This passage clearly shows how the ancients used trees to test the limits of nature, producing strange monsters in the process. Nature, however, would always have the upper hand in the generation game, as it was thought in antiquity

– and well beyond – spontaneously to generate plants and animals. We now turn to that process.

5.2.4 Spontaneous generation

Modern science has refuted the notion of spontaneous generation, both vegetal and animal. Today, we know that having fallen into the soil, seeds may remain dormant for considerable periods of time until conditions suitable for germination are present (and that, as a result, the soil is full of such seeds). We also know that some insects lay extremely small eggs. The ancients had no microscopes and could not observe small seeds or eggs and believed that some plant generation, as animal generation, could be spontaneous.[14] Aristotle explained the phenomenon in the following way in *Generation of Animals*:

> Animals [in particular shellfish] and plants occur in the earth and in the wet because, in the earth, there exists water, and in water there exists *pneuma*, and in all *pneuma* there is soul-heat, so that in a way everything is full of soul. And for that reason they [the animals and plants] are quickly constituted once it [the *pneuma*] has been enclosed. And it gets enclosed when the corporeal (*sōmatikōn*) liquids (*hugrōn*) are heated, and a foamy bubble, as it were, is formed.
>
> (Aristotle, *Generation of Animals* 3.11, 762a19–24)

Pneuma is a complex and changing ancient philosophical notion. Here the term can perhaps best be translated as 'vital breath'. When that vital breath gets trapped in a bubble of 'corporeal liquids' (that is, a mixture of the ancient elements water and earth), and is heated by the sun, a plant or animal can spontaneously grow. Theophrastus explains the phenomenon in a similar way:

> We need to enquire about spontaneous generation. Simply speaking, it must occur when the earth is thoroughly warmed and the accumulated mixture [of earth and liquid] is altered by the sun, as we also observe in the case of animals.
>
> (Theophrastus, *Causes of Plant Phaenomena* 1.5.5)

In other places, Theophrastus insisted on the fact that there must be a qualitative change in the nature of the soil for spontaneous generation to occur. That change in the soil can be brought about by different causes, the most prominent of which is *sēpsis*, decomposition (see e.g. *CP* 1.1.2; 2.9.14).

Theophrastus considered that some plants can be generated both by seed and spontaneously, rather than always spontaneously (*CP* 1.1.2, 1.5.1, 5.4.6). Classicist Otto Regenbogen interpreted this tenet of Theophrastus' thought as scepticism towards the notion of spontaneous generation (1937: 1426). Wilhelm Capelle, on the other hand, stressed the fact that Theophrastus is always prepared to consider different possible explanations for natural phaenomena (1955: 170–171; see also Taub 2003: 124 for Theophrastus' general willingness to consider several explanations). D. M. Balme, for his part, noted that Theophrastus 'is always concerned to give a regular physical cause for spontaneity, and he distinguishes it from marvels' (Balme 1962b: 102). In other words, for Theophrastus the boundary between spontaneous generation and generation by seed was a blurred one. Theophrastus was also aware, unlike many of his predecessors and successors, that some plants, like the cypress, produce tiny seeds and do not grow spontaneously.

Cases of true spontaneous generation, Theophrastus argued, are more common in smaller plants (especially herbaceous ones, *CP* 1.4.6), but may also occur in larger plants. The canonical example of spontaneous generation is that of silphium, the famous plant from Cyrene:

> For it is thus that silphium is said to have come up in Libya, when there had been a fall of rain described as 'pitch-like' and thick, and the forest now existing there is said to have come from another such cause, not having existed before.
>
> (Theophrastus, *Causes of Plant Phaenomena* 1.5.1 (see also *HP* 3.1.6; Pliny, *HN* 16.143))

Theophrastus, and the other botanical authors who discussed silphium, may have failed to appreciate that the seeds of silphium were present in the soil because they were too small to see. If silphium was a member of the Umbelliferae family (see p. 121 for the issues in identifying silphium) it would have possessed tiny seeds, which would have been difficult to see with the naked eye.

According to Athenaeus, Theophrastus also wrote about the generation of truffles:

> Truffles. They occur spontaneously (*automata*) from the earth, especially in sandy places. This is what Theophrastus says about them: 'Truffle (some call it *keraneion*) and any other underground growth'. And again: 'This is the mode of generation and nature of these earth-growths, such as the truffle and the growth that occurs near Cyrene and which they call 'misy'.... About these something peculiar is reported. For they say that when the autumnal rains and severe thunderstorms occur, then these grow, and the more thunderstorms there are, the more they grow, as these are the main cause of their existence.... However some suppose that their origin is through seed (*spermatikēs ousēs tēs archēs*). For on the sea-shore of Mytilene, they say, truffles do not grown until there is a heavy rain and the seed (*sperma*) is brought down from Tiarae.
>
> (Athenaeus, *Deipnosophists* 2.62a–b = Theophrastus *HP* 1.6.9 and fr. 400A Sharples)

As noted by various scholars, apart from the first few words, nowhere do these claims appear in our preserved writings of Theophrastus. On the other hand, a very similar account is found in Pliny the Elder's *Natural History* (19.36–37) in a longer passage on the truffle (*HN* 19.33–37). Athenaeus and Pliny must have used a common source that incorporated Theophrastrean material that is now lost. Regenbogen argues this source is Pamphilus (1940: 1444–1445; see Sharples and Minter 1983; Sharples 1995: 147–149), who may not have quoted Theophrastus particularly accurately. Nevertheless, the juxtaposition of two possible explanations for truffle generation does recall Theophrastus' habit of offering alternative explanations for spontaneous generation. The belief that truffles are generated by lightning was a common one in antiquity, one that founds expressions in the works of the historian-philosopher Plutarch (*Table Talk* 4.2.1, 664b–c) and the poet Juvenal, who describes a lavish dinner in his fifth *Satire*:

> After these [dishes] truffles will be brought in, if it be spring
> And the wished-for thunder-claps our dinner tables
> Enrich. 'Keep your grain for yourself', Alledius said
> 'O Libya, unyoke your oxen, as long as you send us truffles'.
> (Juvenal, *Satire* 5.116–119)

Clearly, postulating spontaneous generation was the only way to explain such amazing phaenomena as the delicious truffle. Humankind could yoke oxen and grow grain that fed the masses, but it could never produce the truffle.

We have now come to the end of our overview of the modes of generation of plants. We now turn to the remainder of the plant's life cycle.

5.3 Growth

Once generation has occurred, plants must grow to survive. We have already encountered some explanations as to how plants grow; we now turn to a fuller discussion of how the Greeks and Romans conceived of this process.

As mentioned above, Greek philosophers compared plant seeds to eggs, as both contained nourishment to feed the developing being. The first to express that idea was apparently the pre-Socratic philosopher Empedocles. Once the nourishment within the seed has been consumed, the plant has to sprout. A description of that process is to be found in the Hippocratic text *Nature of the Child*, where the development of a plant serves as a point of comparison with the development of the human embryo:

> Nourishment and growth of the foetus come from whatever moves from the mother to the womb; and whatever level of health or weakness she may have, the foetus will have the same. In just the same way, things growing in the earth are nourished from the earth, and whatever [characteristics] that earth may possess, the things growing will possess.

> For when a seed is sown in the earth, it is filled with moisture (*ikmados*) from the earth. Indeed the earth contains within itself all sorts of moistures so that it may nourish the things that grow. When the seed is filled with moisture, it swells and inflates. And the power (*dunamis*) in the seed, which is the lightest part of the seed, is forced by the moisture to condense. And as this power condenses under the action of the *pneuma* and the moisture, leaves are formed which shatter the seed.
>
> (Hippocratic Corpus, *Nature of the Child* 22, Loeb 10.60–62 Potter = 7.514 Littré)

For this author, leaves are the first part of the plant to grow, followed by the roots, once the nourishment in the seed is completely exhausted. This particular aspect of the theory seems to have been a matter of debate, as Aristotle argued that the notion whereby 'the formation of the upper part precedes that of the bottom one, also applies to plants. For seeds send out roots before their shoots' (*GA* 2.6, 741b34–36). Now Aristotle's statement seems self-contradictory: the shoots of plants would appear to be the upper part of the plant rather than the bottom one. However, Aristotle is referring here not to the spatial position of plant parts, but to their hierarchical position: the roots are more important to him than the shoots (see Kraak 1942). It is only with the invention of microscopes that cotyledons (seed-leaves) became fully observable, thus making it clear that in plants 'all is leaf'.

Once the plant had exhausted the food in its seed, it had to rely on some other form of nourishment. The notion of plant nourishment, however, posed one conundrum to ancient philosophers. Indeed, nourishment usually involves excretion, but plants were not known to excrete. This is how Aristotle explains the phenomenon:

> Plants take their food from the earth by means of their roots in a treated form. For that reason, plants do not produce residue (*perittōma*). For they use the earth and its heat as if it were a stomach (*koiliai*), whereas almost all animals, and certainly those that move about, have a stomach inside them that is similar to the earth.
>
> (Aristotle, *Parts of Animals* 2.3, 650a21–25)

For Aristotle, the nourishment of plants is already worked-out or concocted by the earth, which acts as a stomach would in animals. Theophrastus, for his part, was reluctant to compare roots to a stomach (*HP* 1.1.11; *CP* 6.11.5, see Chapter 3 for more detail).

Ancient philosophers discussed at length whether parts of plants below and above ground grew at the same rate. Behind that seemingly simple question was a more complicated one regarding the cause responsible for growth in plants. According to Aristotle, the pre-Socratic philosopher Empedocles had attributed the growth of roots to the element earth (which always tends to downward movements), and the growth of aerial parts to the element fire (which always tends to upwards movements). To Aristotle, this made no sense as the plant would end up torn apart under these conflicting movements (*On the Soul* 2.4, 415b28–416a9). Theophrastus too refuted Empedocles:

> For what generates [the plant] is one, and not divided and separated as Empedocles postulates it, attributing the earth to the root, and the air to the shoots, as if each of these [parts] were separated from the other, when in fact they are of the same matter and generated from a single cause.
>
> (Theophrastus, *Causes of Plant Phaenomena* 1.12.5)

That unique cause is the nutritive faculty of the soul – *to threptikon* – an Aristotelian term (see Chapter 3), which Theophrastus only uses once in his works. Theophrastus adds that, while plant growth is caused by a single cause (and not by two as Empedocles had postulated), cold seasons could check the growth of the upper parts of plants. That phenomenon was not, however, caused by the inherent nature of the plant, but by external circumstances. The author of *Nature of the Child* too developed a complex theory of plant growth, in which heat (*to thermon*) plays an important role (see Lonie 1969, 1981). The explanations given by Nicolaus for the phenomenon of plant growth (in book two of *On Plants*) are even more complex, and involve various analogies with other natural phaenomena (such as the salinity of the sea).[15]

Botanical authors also observed that some plants are fast growers, while others are slow. How fast a plant grows depends partly on its species (see Theophrastus, *HP* 3.6.1) and texture. Theophrastus explained that plants with 'a closed texture' grow slowly, whereas those with 'open texture' grow faster, concluding 'this also clearly occurs in animals, and in particular in human beings for the females grow faster than men, being moister and looser in nature' (*CP* 1.8.2 and 4). The rate of plant growth also depends on the nutrition the plant gets, and that nutrition depends on the local environmental conditions (see Chapter 6) and on the 'supplements' the plant got in the form of manure. The ancients described in great detail what type of manure to use in what circumstances (see Figure 5.2), detailing the varying qualities of human, bird and cattle manure (see e.g. Theophrastus *HP* 2.7.4; Cato, *RR* 36; Vergil, *Georgics*, 2.347; Varro; *RR* 1.38; Columella, *RR* 2.13.4; Pliny, *HN* 17.50–57; Palladius, *OA* 1.33; see White 1970: 125–145).

In the right circumstances, a plant could grow through the various stages of childhood, adolescence and maturity. The vocabulary describing plant growth is indeed extremely anthropomorphic. In the case of cultivated trees, ancient authorities often compared the growing process to schooling or education, where the tree was trained and punished when needed (see e.g. Theophrastus, *CP* 3.7.4: *paideia*; see p. 147 for an example in Pliny). Once the plant had reached maturity, it could reproduce, a process to which we now turn, under the heading of annual vegetal processes.

Figure 5.2 Transport of manure (?) on a Gallo-Roman calendar mosaic (third century CE) from St-Romain-en-Gal, France. Saint-Germain-en-Laye, musée d'Archéologie nationale et Domaine national de Saint-Germain-en-Laye 12–553060. Courtesy of RMN-Grand Palais (musée d'Archéologie nationale), photo by Michel Urtado.

5.4 Annual processes: budding, flowering, fruiting and leaf-shedding

The ancients understood that budding and producing flowers was somewhat linked to fruit production. For instance Theophrastus noted that:

> Again, some flowers are sterile (*agona*), as in the case of the flowers that grow at the tips of cucumbers. This is why they remove them [the flowers], for it hinders the sprouting of the cucumber. And they say that

> in the citron, those flowers which have a sort of distaff (*ēlakatēn*) growing in the middle, those are fertile (*gonima*), but those that do not have them are sterile (*agona*).
>
> (Theophrastus, *Enquiry in to Plants* 1.13.4 (on flowers of cucumber, see also pseudo-Aristotle, *Problems* 20.3, 923a13–16))

The 'sterile flower' Theophrastus describes in the case of the cucumber is probably the remains of the flower after pollination and fertilisation have taken place. The cucumber fruit will not develop until these processes have occurred. The 'distaff' that Theophrastus had observed in some citron flowers were probably the pistils of the female flower, that would develop into fruit. Varro, for his part, noticed that, if one plucks the blossom of an unripe fruit tree (a pear), no second one will grow that year 'as the same plant cannot have two periods of pregnancy (*praegnationis*)' (*RR* 1.44.4). Similarly, Pliny noted, in more general terms, the link between flower and fruit:

> When plants flower in truth they give birth (*pariunt*), and the flower makes its appearance with the rupture of the buds (*utriculis*, literally the little womb); the upbringing (*educatio*) takes place at the fruiting stage. This and budding (*germinatio*) are the trees' labour; the flower is the sign of full spring and of the rebirth of the years; the flower is the delight of trees.
>
> (Pliny, *Natural History* 16.94–95)

Pliny did not know that some plants need to be attractive to animals such as bees to be fertilised. In other words, the purpose of the flower as an attracting organ for pollinators was not understood. In fact, the ancients believed the bees 'stole' honey from plants, failing to see that they were fertilising plants (see p. 179 in this book for more detail).

As we saw in Chapter 3, Aristotle – and many after him – considered fruiting to be the *telos* of a plant, its reason for existing. The fruit represented the beginning of the next generation – or at least the potential for a next generation. Plants that did not bear fruits were considered barren, and the more fruit/seed a plant could bear, the more fertile it was judged. Theophrastus devoted much space to the development and maturing of the fruit. He explained that the pericarp (the fleshy part of the seed/fruit, see Chapter 4), developed before the seed *stricto sensu*:

> In all fruits (*karpois*), the pericarp develops before the fruit proper and the seed. This is so not only because many seeds are woody and like a fruit stone, and thus are slower to take form, but also because the seed – as we have said – is the end (*telos*), and what serves an end must exist before that end.
>
> (Theophrastus, *Causes of Plant Phaenomena* 1.21.1)

The Neoplatonist philosopher Porphyry, on the other hand, argued that in the development of a fruit, the principle of that fruit was present from the beginning:

> For just as with a freshly sprouted nut, [at first] everything is together in a continuous fresh green mass, but subsequently its shell and the bony substance under the shell are distinguished, with the tiny membrane under the bony substance and the fruit itself within this membrane as well as the bony [veins] running through the inside of the fruit [all being distinguished] at different times, and yet the nut's form-principle (*logos*) in its entirety is always present in the fruit.
>
> (Porphyry, *To Gaurus*, 17.7, translation: Wilberding 2011: 56)

Porphyry here describes rather precisely a drupe, such as an almond or plum, where there is a woody 'seed' (actually the endocarp, the inside of the fruit wall) inside which is a seed surrounded by a very thin testa.

To ancient philosophers, the process of fruit ripening was one of concoction (see e.g. Aristotle, *Meteorology* 4.2, 380a11–17; Theophrastus, *CP* 2.8.1; pseudo-Aristotle, *On Colours* 5, 795b23–26; *Problems* 20.24–25, 925b24–37), one that affected the odour, flavour and colour of the fruit. Theophrastus explained that this concoction transformed the dry to fluid: it allowed an astringent dry fruit to develop into a juicy one (*CP* 6.6.5). For the Hippocratic author of *Nature of the Child* on the other hand, fruit maturation involved evaporation of liquids:

> But when, over time, the plant has become stronger and its outgrowths have taken root, it acquires broader veins (*phlebas*) running both upwards and downwards. And then it no longer draws from the earth a watery substance, but one that is thicker, more unctuous and more abundant. And when this [substance] is warmed by the sun, it boils up in the extremities, and it becomes fruit according to the lineage of the plant from which it is growing ... When the fruit (*karpos*) has boiled up, it is nourished by the plant, for the plant draws [a substance] from the earth and gives it to the fruit. The sun ripens and strengthens the fruit by drawing out of it the most watery part.
>
> (*Nature of the Child* 22, Loeb 10.64 Potter = 7.516 Littré)

Fruiting was, of course, of high importance from an economic point of view, and barrenness in plants was therefore considered an issue. The ancients had observed that trees producing luxuriant leafage were often unable to produce fruit (see e.g. Theophrastus, *CP* 3.15.4; Columella, *RR* 5.6.36). In Greek, such excessive leafage was referred to as an act of *hubris*, a term that implies a violation of the order of society or nature. Trees that would not perform their duty to society required punishment: the verb *kolouō*, to prune, is linguistically related to the verb *kolazō*, to punish (see Michelini 1978, following the work of Fisher

1976). Conversly, the vocabulary of pruning is sometimes used in a metaphorical sense in some Greek texts. Thus, the Greek-writing Jewish philosopher Philo describes God as 'pruning those branches which through superfluity of nourishment are too luxuriant' (*On Husbandry* 1.6). In the case of the vine, producing too much leafage was referred to as 'going goaty' (Greek: *tragaō*, see e.g. Aristotle, *GA* 1.18, 725b34; Theophrastus, *CP* 1.5.5). Indeed, goats were believed to have a very bad impact on plants in general, and Varro reported that goats were never sacrificed to Athena/Minerva because of the damage goat's spittle caused to the olive tree, which was sacred to the Goddess (*RR* 1.2.19–20, see also Chapter 6 for the damage caused by goats). Other factors that had been observed to impact on the fertility of a plant were age, diseases and accidents (see pp. 151–154), and locality (see Chapter 6). While barrenness was an issue, so was over-production, which would lead the tree to exhaustion and early ageing (see e.g. Columella, *RR* 4.3.4). There too the remedy was pruning.

Besides pruning, the ancients developed other remedies against barrenness. Manure, of course, had a role to play, but more complex recipes were also created. Cato for instance offers the following remedy for a barren olive tree:

> If an olive tree does not bear fruit, turn the earth around it and wrap it with straw. Then mix together equal parts of amurca [i.e. dregs of olive oil] and water and pour it around the olive tree. For a large tree, the amount contained in a water jug should be sufficient; for smaller trees, use a proportionate amount. If you do the same for bearing trees, they will carry even better. But do not wrap these with straw.
>
> (Cato, *On Agriculture* 93)

The ancients noted that wild trees often fail to 'concoct' their fruits to maturity, that is, that their fruits were inedible. Theophrastus explained that this was caused by the dense texture of wild trees, a texture which, as we indicated earlier, would lead to slower – or no – maturation (*CP* 1.15.3, see also Chapter 6 for the difference between wild and cultivated trees). While wild trees were considered as not particularly apt at concocting their fruits, in one particular case, that of the fig, the wild tree (the caprifig, literally, the goat's fig, again indicating the alleged negative impact of goats on fruit trees) helped the cultivated tree to bring its fruits to maturity. It had been observed that some trees, and most especially the fig, drop their fruits before they are ripe. Theophrastus explained that this was caused by a 'certain fluid and *pneuma* caught in the fruit' (*CP* 2.9.3). One had to rid the fruit of this excess fluid, a function played by insects 'produced from wild figs' and hung on the cultivated tree (*CP* 2.9.6, see also Pliny, *HN* 17.255). Theophrastus then described the role of those insects in more detail:

> Upon opening the centre of the fruit, the insects feed on the excess fluid and provide passage from the outside air and, on the whole, make [the fruit] better ventilated. For it happens that, along with the effective

> heat[16] there is, shut up in the fruit, a certain *pneuma* (as in the cases of things that are boiled), and when this has been removed and transformed into vapour together with the excess fluid, the fruit remains on the tree.
>
> (Theophrastus, *Causes of Plant Phaenomena* 2.9.6 (see also Pliny, *HN* 15.79))

A bit further on, Theophrastus notes that insects are not needed in dusty places, as that dust will dry out the fruits (*CP* 2.9.7, see also Pliny, *HN* 15.79–81). In terms of modern botany, the fig is dioecious: male and female flowers are found on separate trees. Cultivated fig trees are female and wild fig trees – or caprifigs – are male. The pollen from the male caprifigs needs to be carried to the female cultivated figs. That function is performed by a wasp (*Blastophaga psenes* L.), which acts as a pollinator. That wasp does not 'ventilate' the fruit but fertilises it. The dust that Theophrastus had observed in some cases of fig fructification was of course pollen carried by the wind.

Greek and Roman authors, starting with Herodotus (who gets his facts quite wrong, *Histories* 1.193; Theophrastus *HP* 2.8.4; *CP* 2.9.15; see Georgi 1982), noted the similarity between what happens to the fig in the process of caprification and what happens to the date palm when 'male' spathes are shaken over the 'female' tree. From a modern perspective, these are both pollination processes. However, while as we saw earlier, ancient authors described the love of the palms in sexualised terms, they did not do so in the case of caprification. They simply did not know that the phenomenon they were observing was one of fertilisation.

We have now discussed the annual processes of flowering and fruiting, which are linked to generation. Before we turn to the end of the life of a plant – its old age and death – we must say a few words about another annual process: leaf-shedding. A distinction was made in antiquity between deciduous and evergreen trees. Theophrastus explained that trees are evergreen because of the continuous supply of nourishment, adding that 'some are [evergreen] on account of their nature, and others on account of their region' (*CP* 1.11.6). Deciduous trees assist evergreen ones by providing them with 'manure' with their decomposing leaves (*CP* 2.18.1).

Both Aristotle and Theophrastus drew an analogy between leaf-shedding in trees and horn-shedding in stags, feather-shedding in birds, and hair-shedding in four-footed animals (Aristotle, *GA* 5.3, 783b2–9; Theophrastus, *HP* 1.1.3). In the *Meteorology*, Aristotle quotes the following verses of Empedocles: 'The same are hair and leaves and thick feathers of large birds, and scales upon sturdy limbs' (*Meteorology* 4.9, 387b5–6 = Empedocles DK 31B82). Aristotle explained leaf-shedding (and all other sheddings) as a heat deficiency, one that occurred every autumn/winter in deciduous plants living in temperate regions. Pliny added that the environmental conditions had an influence on how late or early a tree shed its leaves (*HN* 16.82).

5.5 Old age and death

As Pliny noted, in old age, plants bear more and faster, their bark becomes more wrinkled, and they give out stronger scent (see e.g. *HN* 16.116 and 126; 21.36). The ancients also observed that some species of trees grow old faster than others, for instance the pomegranate, laurel and apple tree (see e.g. Pliny, *HN* 16.241). Conversely, they knew that some trees – in particular wild ones – are very slow to age (e.g. the oak, chestnut or walnut) or appear not to age at all (e.g. cypress, cedar, ebony); and some grow to an incredible age (see e.g. Theophrastus, *HP* 4.13; Pliny, *HN* 16.212). In fact, Aristotle, noted that the longest-lived beings are to be found among plants (*On Length and Shortness of Life* 4, 466a10). He explained that life is moist and warm, while old age is cold and dry:

> It is among plants that the longest-lived beings are to be found, more so than among animals. First, because they are less watery, so that they are not easily congealed. Second, they have fattiness and stickiness, and while being dry and earthy, they have a moisture which is not easily dried. But we must consider why trees are so long-lived by nature. For in comparison to animals (apart from insects), plants have their particular [nature]. Plants are always reborn, and for that reason they are long-lived. For there always are new suckers, while others grow old. And the same goes for the roots.... As was mentioned above, plants resemble insects; for they live even when divided, and two or more grow from one.
>
> (Aristotle, *On Length and Shortness of Life* 6, 467a7–20)

The longevity of plants, then, is due to the texture of trees (an oily texture) and to the ability of trees to produce suckers and thus renew themselves.

Despite the longevity of some plants, the life of a plant must end with death, whose cause could be either natural or unnatural according to Theophrastus. Natural causes included old age and weakness. Unnatural causes included being chopped, killed by adverse weather, by excessive fruit-production, by other plants (such as ivy), and by disease (*CP* 5.11.1). This classification of the causes of death into 'natural' (*kata phusin*) and 'unnatural' (*para phusin*) is the same as the one found in the medical treatises of the time.

The *Cesti*, an encyclopaedic collection of facts attributed to Sextus Julius Africanus (second–third century CE), contains interesting information on how to kill trees in a military context, to deprive the enemy of food. Some tips are simple, such as spreading salt in the fields. Others are more colourful:

> At all events you will dry up all trees, with the exception of the apple, if you stick the sting of the sting ray into the trunk of the tree. The Quintilii [two brothers who lived in the second century CE, see Rodgers

> 2008e for references] say that the pods of beans thrown against their roots dry them up.
>
> (Sextius Julius Africanus, *Cesti* 2.3)

In the past as today, human beings could be the worst enemy of plants, and could most definitely end the life of the longest-lived tree. We have now reached the end of the life cycle of a plant, but it remains to examine the diseases of plants, diseases that could affect them at any age.

5.6 Diseases

Plants, like animals, suffer diseases that can lead to sterility, weakening or death (see Strömberg 1937: 187–195). Theophrastus divided diseases into natural and unnatural, noting that:

> Diseases on the whole would appear to be unnatural (*para phusin*), since disease is always a deviation from and a violation of what is natural (*kata phusin*); but because they are customary and occur regularly, we say that they are natural both in animals and plants, distinguishing however from the rest, the states that are caused by violence and which manifestly result from external causes, such as wounds and blows.
>
> (Theophrastus, *Causes of Plant Phaenomena* 5.8.1)

This distinction between natural and unnatural again is very similar to the classification of human diseases current at the time. Theophrastus divided the causes of diseases into internal (too much or too little food) and external. The external causes, in turn, he divided into natural (temperature or quality of the air) and unnatural (blow or wound). Theophrastus reports that certain (unnamed) authorities believed all diseases arose from blows and wounds, but he expresses his disagreement with this theory (*CP* 5.9.6; *HP* 4.14.7). The boundary between external and internal causes is blurred in the case of worms, which in some cases are said to be born spontaneously from a plant (see e.g. *HP* 4.14.2 and 5).

Theophrastus also divided diseases into those of plants and those of fruits, which include grains. Some diseases are common to all plants, while others are specific to some types of plants; some are common to all types of 'fruits' while others are particular to some types of fruit-producing plants (*CP* 5.8.2–5.10.5). Pliny adopts a similar classification of plant diseases, discussing diseases of trees in book seventeen, diseases of cereals in book eighteen and diseases of garden vegetables in book nineteen (*HN* 17.216–240, 18.149–156, 19.176–177).

As Pliny observed, plants suffered diseases 'sharing even their names with the sufferings of humankind'. Indeed, they were affected with, among other things, pains in the limbs, indigestion and obesity (*HN* 17.219). In the case of diseases as in other aspects of plant life, Pliny exaggerated the comparison between plants and animals and humans: 'Sometimes [trees] die from contagious diseases according to their kind, just as among humans, diseases sometimes attack slaves, and sometimes the urban or rural plebs' (*HN* 17.219).

Many of the tree 'diseases' described by ancient authors are in fact fungal growths, but these growths were not recognised as being independent organisms. Thus the 'cobweb' that affects the olive (Theophrastus, *HP* 4.14.10; *CP* 5.10.2); the scab that affects the fig; and the *krados* that affect the same tree (*CP* 5.9.12; *HP* 4.14.5) are all fungal mycelia. The ancients had observed that these 'diseases' are more likely to occur in certain climatic conditions. For instance, Theophrastus noted that some give this account of the causes of scab:

> Some believe that scab occurs in another way [rather than through over-nourishment], for instance when there is light rain at the rising of the Pleiades. For then [the fig trees] ferment and, being heated, they send out [the fermenting agents?] to the outside, like pustules. But if [the rain] is heavy, the causes [of the fermentation] are washed off, or [the trees] transmit them to their inside, where they stop.
>
> (Theophrastus, *Causes of Plant Phaenomena* 5.9.12)

The 'pustules' observed by Theophrastus' sources are fungal mycelia. However, these sources could not have come to that conclusion, even though they had observed the conditions which promote fungal growth and hence cause unhealthy trees. Theophrastus had also observed that some species, and in particular wild species, are less prone to diseases than others. For instance, he noted that the wild fig did not suffer from *krados* rot or scab (*HP* 4.14.4). It is a well-established fact that some species of plants, and in particular, certain strains of cultivated plants, are more susceptible to fungal diseases than others.

Ancient remedies for plant diseases have also been transmitted to us. These include familiar treatments such as manure, pruning, using ashes, salt or brine. In Pliny's writings some of these treatments are compared to treatments applied to humans. For instance, he talks of 'surgical bloodletting', and of pruning he writes that 'to a large extent the medical treatment of trees resembles that of human beings, as the bones of trees also are treated by perforation' (*HN* 17.246 and 17.252). While Theophrastus does not mention any remedies for grain diseases, the Roman agronomists and Pliny transmit some rather interesting ones. For instance, these are the remedies Columella recommends to avoid worms:

> Democritus in his book, which in Greek is entitled *On Antipathies*, maintains that these same little beasts are killed if a woman, who is menstruating, walks three times around each planting area with her hair loose and her feet bare; for after this all the little worms fall down and thus are killed.
>
> (Columella, *On Agriculture* 11.3.64 (see also Pliny, *HN* 18.158)[17])

As we mentioned in Chapter 2, the name of the pre-Socratic philosopher Democritus was attached to various magical treatises. Using the power of a menstruating woman in a state of undress was a magical ritual. In ancient economies, pest infestations could have terrible consequences, and at times magic was the only recourse – prevention was of course better than cure.

5.7 Conclusions

We opened this chapter with verses from a botanical poem. We close them with other verses, from the Latin collection of poems known as the *Priapeia*, which celebrate the phallic god Priapus, whose simple wooden statues protected gardens:

> I am not hewed from the fragile elm,
> Nor is that which stands stretched out with rigid vein –
> This column of mine – made out of any wood,
> But generated from the evergreen cypress,
> Which neither the generations a hundred-times completed
> Does fear, nor the decay brought by lengthy senility.
> You, whoever you may be, O evil one (*malus*), fear this
> For, if with greedy hands even the smallest
> Clusters from this vine-sprout you damage,
> There will be born, however much you may want to refuse it,
> On you a fig grafted (*inserta*) from this cypress.
>
> (*Priapeia* 91)

The threat at the end of the poem is a sexual one: 'fig' (Latin: *ficus*) was a name for the male sexual organs (see Buckheit 1960), and the word we translated as 'grafted' – *inserta* – can also be translated as 'thrust in'. In other words, the god Priapus is threatening to bugger whoever damages his statue. There are many other playful puns in the poem: for instance the word to address the malevolent passer-by – *malus* – is also the name of the apple tree (for this type of humour, see Richlin 1992; Uden 2010). This poem also includes many well-informed references to themes which we have studied in this chapter: the masculinity of hard wood; the generation of plants; the process of grafting; and the fact that some trees age very slowly. This is a lot to pack in eleven verses! Greek and Roman authors who would not have identified themselves as plant experts often displayed excellent knowledge of plant physiology, and they fully exploited the metaphorical potential of plant physiological processes. Conversely, ancient botanical texts are often filled with allusions to the life of humans and to human or animal physiology. Thus we have encountered in this chapter plants in love, plants getting married, plants adopting others and giving birth to their kin, plants suffering the same ills as humans, and dying the same death. Ancient authors often used to describe plants vocabulary that had been created to describe human families, societies and physiological processes. They did so because analogy is one of the best ways to understand the mysteries of nature; they also did so – in particular in the case of Pliny – because they wanted to entertain their audiences. Readers could enjoy texts about the life of plants on several levels: they could learn about physiological phaenomena, but they could also smile or cry at the similarities between the fates of humans and plants. For plants, which served humans in so many ways, often had the last laugh in nature: wild majestic trees outlived the longest-lived of animals, and the seemingly weaker herbaceous plants had the incredible power to sprout as if produced by rain. Plant vitality surpassed that of any animal, even the most rational animal: man.

Notes

1 On anthropomorphism and anthropocentrism in ancient botany, see Amigues 1999a: 149–154.
2 There are many other passages where Aristotle discusses plant generation. See Totelin forthcoming (b) for an introduction and discussion of the links between animal and plant generation in Greek and Roman texts. On male and female principles, see Tortzen 1991; Negbi 1995.
3 On metaphors of the female body, see DuBois 1988; King 1994.
4 On the question, see Wöhrle 1985: 53–62; Tortzen 1991; Negbi 1995; Foxhall 1998; Bretin-Chabrol and Leduc 2009. This gendering of plants continued throughout the Middle Ages, see Cadden 1995: 207–210.
5 It was a Greek trope to see the Egyptians as doing everything in the opposite way to the Greeks.
6 The literature on this topic is large. See in particular King 1998: chapter one and Mayhew 2010: 98–99.
7 Palladius (*OA* 11.12.3) also notes that planting pistachios in male and female pairs brings better results. The case of the pistachio, however, never attracted as much attention as that of the palm tree.
8 On these planting pots, see Thompson 1937; Jashemski 1974 (Pompeii); Detsicas 1981 (Eccles); Messineo 1984; Jashemski 1992; Macaulay-Lewis 2006a and 2006b (Petra).
9 For nurseries discovered archaeologically, see Kenawi *et al.* 2012.
10 Note that in this paragraph of Cato, the slips are referred to as 'seedlings', *semina*, even though it is clear that most trees grown in nurseries would not have been grown from seed.
11 On ancient grafting, see White 1970: 248–258; Pease 1933; Ross 1980; Clément-Tarantino 2006; Bovey 1999; Lowe 2010; Thibodeau 2011: 145–150.
12 Ross 1980 would translate this sentence as follows: 'For example, an oak does not take the graft of a pear, nor, even if it is an apple, does it take the graft of a pear.' He argues for this translation on the basis that, in his belief, pears cannot be grafted onto apples. However, that graft is possible.
13 Thibodeau (2011: 147) talks of 'space parasites'.
14 On animal spontaneity in Aristotle, see e.g. Balme 1962b; Lennox 1981; Gotthelf 1989; French 1994: 65–69; Lloyd 1996: 104–125. For animal spontaneous generation in Theophrastus' treatise *On Fish*, see Sharples 1992.
15 Our knowledge of Arabic (which is the language of the best version of *On Plants*) is too limited to offer any useful comments here, and research on Nicolaus' theories or growth deserve further study.
16 On the notion of heat in Theophrastus and his predecessors, see Viano 1992.
17 On ancient 'pesticides', see Smith and Secoy 1975.

6

AIRS, WATERS AND PLACES

Plants and their environments in antiquity

6.1 Introduction

The Greeks (and the Romans after them) were travellers; they explored the Mediterranean, its hinterland, and beyond. Through their travels and conquests, they discovered 'new' peoples, animals and plants; they observed familiar animals and plants in 'new' environments; they brought back human slaves, animals and plants. They knew that even the furthest reaches of the earth, with their unwelcoming climates (by Greek and Roman standards) grew useful plants. Pliny the Elder, in particular, stresses the link between botanical knowledge and empire. In the following passage, in addition to making imperialist claims, he shows that plants grow in incredibly varied ecological conditions:

> [We get] the Scythian herb from the marshes of Maeotis [i.e. regions surrounding the Sea of Azov; see Map 3]; euphorbia spurge from Mount Atlas and from beyond the Pillars of Hercules [i.e. the straights of Gibraltar; see Map 3] – that place deserted by the works of Nature; from another part [of the earth] the plant *britannica*, from islands in the Ocean situated beyond the mainland; the plant *aethiopis* from a clime burnt by the stars; and we bring plants from other places throughout the whole world, backwards and forwards, for the benefit of humankind, all thanks to the immeasurable grandeur of the Roman Peace, which shows us not only humans different from each other in their lands and tribes, but also mountains and peaks married to clouds, as well as their offspring, and again herbs.
>
> (Pliny, *Natural History* 27.2–3)

Book twenty-seven of Pliny's *Natural History* is a book devoted to medicinal plants, and it is therefore not surprising that Pliny should put such emphasis on 'exotic' plants here. However, strange plants found in a particular locality interested many authors beside medical writers and encyclopaedists. Geography or travel writers, such as Strabo (first century BCE–first century CE) and Pausanias (second century CE), described in some detail the flora of the places they discussed, whether they had visited them or not (see p. 55 in this book). Accounts of semi-fictional travels, such as those of the wise man Apollonius of Tyana (first century CE) to the East, as recounted by Philostratus (second–third century CE), include many descriptions of plants, as in the following example describing 'Indian' trees:

> On the steep crags of this mountain, there grow lofty frankincense trees and many other types of trees, as for instance the pepper tree, which monkeys cultivate. And they did not neglect to describe this plant, and I shall depict it in the same way as the account was given. The pepper tree resembles the Greek chaste tree in all respects and in particular in the clusters formed by its fruits. It grows in remote places that are not easily accessible to man, and a community of monkeys is reported to dwell in the recesses of the mountains, in some of its caves. And the Indians hold these monkeys in great esteem, because they pluck the 'grapes' and they drive away the lions with dogs and weapons.
>
> (Philostratus, *Life of Apollonius* 3.4)

Greek and Roman writers knew that there is a strong link between humans, animals, plants and their environment. That link, however, was especially strong in the case of plants, as they are bound to the earth through their roots, as expressed by Theophrastus, in the *Enquiry into Plants*:

> It is also fitting to add [to the list of *diaphorai* between plants] the places in which each type is able to grow or not. For this is a major difference and one that, above all, is particular to plants. Indeed, plants are bound to the earth and not independent from it as animals are.
>
> (Theophrastus, *Enquiry into Plants* 1.4.4)

The philosopher also noted that plants seek (*zētei*) the right environment and climate for themselves, and when they have found those perfect conditions they grow most beautiful:

> So far, we have examined the differences between plants of the same type. But all plants become more beautiful (*kalliō*) and healthier in their proper environment (*oikeois topois*).... Each plant seeks (*zētei*) its proper environment and climate. The proof: some places bear some plants but not others, which will neither occur of their own accord there nor grow easily; and when these plants eventually take hold, they do not bear fruits.
>
> (Theophrastus, *Enquiry into Plants* 4.1.1 and 5 (see also *CP* 2.16.8 for the same use of *zētei*))

The Greek words to designate the 'proper environment' for a plant are '*oikeios topos*', literally 'the homely place'. Greek *oikos* means the house, its inhabitants, and the estate that is attached to it – the *oikos* was the basis of ancient Greek economic life. Modern words that are derived from the Greek *oikos* include 'economy' and 'ecology', both disciplines that are still relevant to a study of botany. The science of ecology is a modern one, with its roots in the eighteenth and nineteenth centuries (see Hughes 1988). Theophrastus and other ancient authors, however, were aware of the interaction between plants and their environment. As environmental historian Donald Hughes pointed out, Theophrastus was not advancing a theory of environmental determinism, in which external factors completely control the expression of characteristics in an organism; rather the philosopher stressed the interaction of environmental factors with the plant's inherent tendencies (Hughes 1988: 69).

The same teleological insistence on the beauty of plants growing in the right environment is found in various passages of the *Enquiry*, either in general terms (see e.g. *HP* 1.9.2) or in particular examples, as in the following one: 'While the [terebinth] tree is short, shrub-like and twisted around the Ida and in Macedonia; around Damascus in Syria [see Map 3] it grows tall, abundant and beautiful' (Theophrastus, *Enquiry into Plants* 3.15.3). In fourth-century Athens, where Theophrastus taught philosophy, the notion of 'beauty' encompassed more than physical appearance, to include moral qualities. A noble man was *kalos kagathos*, good and beautiful. We would argue that, when Theophrastus compares the beautiful Syrian terebinth to the twisted Macedonian one, he is imparting to the tree the same moral attributes of beauty and turpitude as those he would give to men in his *Characters*. Once again anthropomorphism is ever-present in ancient botanical writings.

The Greeks and Romans recognised that what constitutes 'the proper environment' varies from one plant to the next. While some plants sought rich soils and temperate climes, others thrived in rocky grounds on windy hill tops. Some plants only grew well under the care of men; others only grew wild. Anyone who made a living out of plants – and there were many in the ancient world (see Chapter 2) – had to be aware of those facts. Farmers, in particular, had to know which part of their farm would be best to grow which type of plant, as Scrofa (one of the interlocutors in Varro's *On Agriculture*) makes it clear:

> Two things are to be considered: what is most advantageous to sow and in what place. For some places are suited to hay, others to grain, others to vines, others to olive, others to crops that are suitable as fodder.... It is not right to sow every kind of plant in rich soil, nor to sow nothing in lean soil. For it is better to plant in leaner soil those plants which do not require much sap.
>
> (Varro, *On Agriculture* 1.23.1–2)

Beside the soil, consideration had to be given to winds and their directions (see e.g. Theophrastus *CP* 2.3.1–3; Pliny, *HN* 17.10, 18 and 24) and to the quality of the water (see e.g. Theophrastus *CP* 2.5.1–2.6.4).

While climate and soil could not really be altered, ancient agronomists trusted that through agronomical art even the worse ecological conditions could be improved. This is what Ennius, another of Varro's interlocutors, has to say on the topic:

> For, even though salubrity, which is a function of the climate and soil, is not in our power, but rather in that of Nature, it still depends greatly on us, since through care we can lessen its worst effects. For if a farm is really unhealthy on account of the soil or the water, from the stench which is exhaled in a certain spot; or on account of the climate the estate is too hot or the wind that blows is not healthy – these issues can be improved by the science and investment of the master.... Did not the physician Hippocrates himself during a great pestilence save not one estate but many cities with his science?
>
> (Varro, *On Agriculture* 1.4.4–5)

Varro's Ennius is here referring to the legend whereby Hippocrates or his students cured 'plague', as described in the pseudepigraphic Hippocratic *Letters* (25, p. 106 Smith = 9.400 Littré). While it is unknown whether Hippocrates ever cured people suffering from 'plagues', it is certain that texts of the Hippocratic Corpus postulate a link between health and environment. The best known of these texts is *Airs, Waters and Places* (Loeb 1.65–137 Jones = 2.1–93 Littré), which Theophrastus may well have known – Aristotle certainly seems to have been aware of the ideas contained therein (see Byl 1980; Oser-Grote 2004).

Throughout this book, we have noted various links between plants and their environment. Here we look into more detail at some of these interactions. We start with the most basic environmental distinction: that between wild and cultivated plants, which were considered to be of different 'types' in the ancient world. As examples of cultivated and wild environments respectively, we focus more particularly on gardens and mountain forests. We then turn to plants identified by geographical names in ancient texts (e.g. Illyrian iris), the transplantation of foreign plants in Greece and Rome, and the issues the ancients encountered in attempting to bend plants to new climates. We then briefly examine aquatic plants, and more particularly the algae described in ancient botanical texts. We end this chapter with some considerations on plant extinction and climate changes in the ancient world.

6.2 Cultivated and wild; garden and forest

The ancients often differentiated between wild (that is, unaided by agriculture) and cultivated types of plants. For instance, they considered the wild fig to be a different type (*genos* or *eidos*) of plant from the cultivated one (see e.g. Theophrastus *CP* 1.18.3; see French 1994: 96–98). Modern botanists do recognise that plants develop differently in wild and cultivated environments. However, to them, a fig is a fig (*Ficus* spp.): the wild variety is not a different species from the cultivated variety. To ancient botanical authors, on the other hand, the distinction between wild and cultivated was crucial. For Theophrastus that distinction was the second most important classificatory criterion after plant parts. For instance, book two of the *Enquiry* is devoted to cultivated trees and shrubs; whereas book three is devoted to wild trees and shrubs; and similar divisions between cultivated and wild are found throughout the *Enquiry* (see Chapter 3). In *Causes of Plant Phaenomena*, Theophrastus takes a more theoretical approach to the topic, and asks whether the growth of cultivated and wild plants is equally natural (*kata phusin*). He first considers the case that only 'spontaneous' (that is, unaided by agriculture) growth is natural, but then argues that:

> For fulfilment (*teleiōsis*) of a plant's nature also occurs when it obtains from the art (*technē*) whatever it happens to lack, such as the right quality and quantity of food and the removal of hindrances and impediments, all of which are clearly provided by the regions appropriate (*oikeioi topoi*) to each plant – it is in those places indeed, we argue, that one should observe the nature of these plants. But these regions only provide external [help], such as air, wind, soil and food, while agriculture also alters and changes them from within.
>
> (Theophrastus, *Causes of Plant Phaenomena* 1.16.11–12)

He concludes that some plants, by nature, grow better without human aid, whereas others need the benefits of agriculture – a few grow well in both circumstances (*CP* 1.16.13). In other words, human art helps nature to achieve the fulfilment, the *telos*, of plants.

The division between cultivated and wild is also important in the works of the Roman agronomists, who – quite naturally – were more interested in agricultural products than in wild plants. This is how Columella presents the distinction at the very beginning of *On Trees*:

> It seems right to us, as it does to Vergil, to divide young shoots into two types (*genera*): those that grow of their own accord, and those that proceed from human care. The former, which does not result from human work, is better suited for timber; the latter, on which effort is expended, is suitable for [growing] fruits.
>
> (Columella, *On Trees* 1)

Cultivated plants then are especially important as they provide humans with food. According to the ancients, wild plants were not – on average – as wholesome as cultivated ones – in the words of Galen: 'thus they count these among the wild plants, and in general, it is sufficient to know that they all are full of bad juices (*kakochuma*)' (*Foodstuffs* 2.39, 6.624 Kühn). In Theophrastean terms, wild plants do not have the ability to concoct their fruits to the point where they become properly edible. Theophrastus concludes his discussion of why wild trees do not produce fruits that are edible for humans as follows:

> There is certainly a concoction of the pericarp, but there is also one of the fruit proper; and the former [concoction] provides our food, while the latter serves the generation and continuation of the tree, for fruits and seeds have these purposes.
>
> (Theophrastus, *Causes of Plant Phaenomena* 1.16.1)

Although wild plants, on the whole, did not provide wholesome food for humans, they were by no means useless. They provided timber; indeed, some useful timber trees such as the oak tend to deteriorate when planted by man (see e.g. Theophrastus, *HP* 2.2.6; on timber in the ancient world, see Meiggs 1982). In addition, most ancient medicinal plants were wild plants, which only deteriorated under human cultivation:

> Hence it is reasonable that husbandry should not be beneficial to certain plants, such as those that are dry, pungent and bitter, and put simply, those that are drug-like (*pharmakōdē*) and useful to us in this manner. For they are made effeminate (*ekthēlunetai*) when their powers are removed and some do not even carry fruit at all, others [carry] fruits that are watery and inferior, and the plants themselves become more watery, as in the case of wormwood, centaury, and in general all plants with drug-like properties, whether those are to be found in the root, the fruit or the branches.
>
> (Theophrastus, *Causes of Plant Phaenomena* 3.1.3–4 (see Totelin 2015, with the same translation))

Significantly, pharmacological plants that are grown under cultivation become 'effeminate'; they lose their masculine strength. Pliny too notes that saffron (a plant used as spice and medicine in the ancient world) may look more beautiful when cultivated, but is in fact much weaker: 'wild saffron is the best ... The cultivated saffron is broader, larger and more beautiful (*nitidius*), but much milder' (*HN* 21.31). The 'beauty' of plants that grow in the right environment is more than 'skin deep': a weak plant may look more beautiful than a strong one, but it lacks real beauty.

The ancients wondered why some plants, such as the caper or silphium (two plants used both as medicine and food) refused to grow under cultivation. Here is how the pseudo-Aristotelian *Problems* answer the question 'why does the caper tree not want to grow easily in cultivated places?'

> For many have attempted to transplant roots or to sow seeds (for in some places they fetch higher prices than roses), but they grow best on tombs because the ground is untrodden.... Farming concocts and makes

> productive the nourishment: from this, cultivated fruits are formed. Thus, the plants that grow from this cultivation are called 'cultivated' because they are helped by expertise (*technē*), as if they were taught (*paideuomena*). But those that cannot [be cultivated] and which are formed out in opposite conditions, these are wild and do not want (*ethelei*) to grow on cultivated ground. For farming destroys (*phtheirei*) these by teaching them, since they come out of corruption [i.e. corpses, *phthoras*].
>
> (pseudo-Aristotle, *Problems* 20.12, 924a1–23 (see also e.g. Theophrastus, *CP* 3.1.4–6))

In this explanation, farming concocts (perfects) plants, but when a plant comes from rot (that is, grows on tombs) cultivation will destroy it by going against the very nature of the plant. Caper was one of the plants that was wild by nature, and which could never be submitted to cultivation.

Cultivated plants were not perfect in all respects. Rather, as the vocabulary used in the passage from the pseudo-Aristotelian *Problems* shows, they were like children in need of education – a time-consuming process. These plant-children also tended to consume much more 'food' than their wild relatives, thus leading to fears that Mother Earth might exhaust herself. Columella repeatedly criticised those 'leaders of our state' who thought the earth had been exhausted by over-production, like an old woman exhausted by numerous pregnancies (*RR* preface 1; 2.1.2–3). In Columella's opinion, however, nature was endowed with constant fertility. If crops failed, it was because of humans' lack of skill and knowledge. To obtain from the earth constant returns, one must have a full understanding of soils, winds, waters, and the nature of plants.

We have discussed in general terms the differences perceived between cultivated and wild plants in the ancient world. We now turn to some examples of cultivated and wild environments. There were innumerable environments in which plants grew. Cato, for instance, lists nine cultivated environments in his description of the ideal farm:

> If you should ask me what type of farm is best, I would say: one hundred iugera of land, comprising all sorts of soils, and in a good location: a vineyard is first, if it gives good and plentiful wine; second an irrigated garden (*hortus*); third a willow-grove; fourth, an oliveyard; fifth, a meadow; sixth, cereal land; seventh, woodland for cutting; eighth, an orchard; and ninth, a mast-grove.
>
> (Cato, *On Agriculture* 1.1.7)

It is impossible to describe all plant environments here. Instead, we shall focus on the garden and the mountain forest, as the representatives, respectively, of the cultivated and wild environments.[1]

6.2.1 The garden and its plants

'Garden' (see Figure 6.1) is a vague term, and can cover a wide range of realities: from the so-called 'gardens of Adonis' (a few seeds sown on a potsherd at the hottest time of the year in honour of the deity Adonis, see Detienne 1972), to tomb gardens (see Jashemski 1970; Brundrett 2010), private vegetable or fruit gardens, private ornamental gardens, temple gardens/groves (see Dillon 1997),

Figure 6.1 Representation of the mythical Garden of the Hesperides on a red-figure Attic lekythos (*c.*420–390 BCE). Attributed to the Circle of Meidias Painter. Getty Villa 91.AE.9. Courtesy of the J. Paul Getty Trust, California.

and enormous palace gardens. In Greek these are all *kēpoi* (singular *kēpos*), with the exception of palace gardens or parks, that are referred to as *paradeisoi* (singular *paradeisos*), paradises (see Bremmer 1999). In Latin too, the word *hortus* is polyvalent, as noted by Pliny:

> In our laws of the XII Tables, nowhere is the word '*villa*' [i.e. farming estate] used; it is always the word '*hortus*' that has that meaning, while the word '*heredium*' [which usually denoted a family estate] has the meaning of 'garden' ... Today, some people possess under the name *hortus* the pleasures of farms and country houses within the city itself.
> (Pliny, *Natural History* 19.50–51)

Ancient gardens, and in particular Roman ones, have attracted much scholarly attention in recent years, both from archaeologists and historians working with textual or material evidence.[2] As noted in Chapter 1, garden archaeology has become a fruitful field of research since the pioneering work of Wilhelmina Jashemski at the Vesuvian sites. However, there are several issues here. First, while gardens of the Roman world are now well known, far less work has been done on the Greek world. Second, the most difficult thing to recover through garden archaeology is plants. Pollen is not always available for analysis; and while the cavities left by the roots of a plant can indicate what type of plant grew there, it cannot give definite answers as to the identity of the plant. Certainty is only possible when a seed is found near a tree cavity, as in the case of an apricot kernel found near a tree cavity in the garden of the House of Bacchus and Ariadne at Thuburdo Maius (modern Tunisia, see Map 3, see Jashemski *et al.* 1995: 564). Third, it is difficult to use textual or material evidence to 'fill in the gaps' in the archaeological evidence. Texts (literary or documentary) and artefacts (paintings, mosaics, etc.) have their own conventions and agendas.

Despite these difficulties, garden archaeologists and other scholars working on gardens in recent years have come to similar conclusions regarding the importance of gardens in social competition. Through their gardens, the ancients displayed their wealth and social status. Kings, generals, emperors, and other important political figures used gardens as symbols of their power. Poets and artists celebrated gardens in such terms. By the first century CE, Roman gardens had become places of luxurious display. According to Pliny the Elder, who gives a short history of gardens in book nineteen of the *Natural History*, this phenomenon was an import from Greece, where the philosopher Epicurus had started a fashion for luxurious gardens (*HN* 19.51). To Pliny, these luxurious ornamental gardens were problematic because they robbed the poor of food, while promoting useless produce, such as fruits grown to extraordinary sizes, vines whose wine had to be matured for more than a lifetime, and a cabbage 'fattened to such a size that the table of a poor man could not accommodate it' (*HN* 19.54). The poet Martial, too, deplored useless ornamental gardens in his praise of Faustinus' farm:

> The Baian villa, dear Bassus, of our friend Faustinus
> Is not ordered with idle myrtle-groves,
> Widowed plane trees, and clipped box-rows,
> It does not occupy an unpleasant expanse of broad field,
> But in a true uncultivated (*barbaro*) estate it does rejoice.
> (Martial, *Epigrams* 3.58.1–5)

Faustinus' farm concentrates its energy on the production of corn and wine and the rearing of animals. Later in the poem, Martial also praises Faustinus' refined garden (*hortus urbanus*, verse 29). By contrast, Bassus' garden is barren:

> But you possess in the suburbs elegant hunger
> And from lofty towers, you gaze upon nothing but laurels,
> Here Priapus fears no thief; you are secure.
> (Martial, *Epigrams* 3.58.45–47)

Instead of calling Bassus' garden a *hortus*, he names it by metonymy 'elegant hunger'. This garden may be beautiful, but it is useless, and there is nothing to steal growing there. It is one of those gardens where 'widowed plane trees, and clipped box-rows' thrive. Martial is here alluding to the topiary art, which was allegedly brought to perfection under the rule of the Emperor Augustus by a member of the Equestrian order, Gaius Matius (Pliny, *HN* 12.13).

In the preface to book ten of *On Agriculture* – the book devoted to gardens, which Columella chose to write in verse rather than in prose – Columella also stresses the fact that gardens have become the place where wealth is competitively displayed, but he hints at more productive functions. After praising the merits of a golden age when even the poor (*pauperibus*) enjoyed a good diet, he describes the current situation:

> Soon, as subsequent ages, and particularly our own, placed a fanciful value on meals, and dinners were no longer considered as fulfilling natural desires, but as means to display wealth, plebeian poverty, driven away from costlier dishes, has been compelled to consume ordinary ones. Therefore the cultivation of gardens, since their produce is now in greater use, requires from us more careful instruction than what our elders have handed down.
> (Columella, *On Agriculture*, preface to book 10, 2–3)

For Columella, gardens and garden-writing are particularly important in this age where the gap between the richest and the poorest is ever increasing. While we cannot deny that there was social competition located in the garden, we also note that the primary function of ancient gardens can sometimes get lost in the narrative of display and power. That primary function was of course an economic one: it is in gardens that much useful produce was grown. The ancients often named a cultivated variety of plants *kēpaios* (Greek)/*hortulanus* (Latin), literally 'of the garden'. This epithet is commonly found in ancient herbals, which describe the pharmacological powers of plants. For instance, Dioscorides devotes one chapter to the 'garden celery' (Greek: *selinon kēpaion*, our celery, *Apium graveolens* L.) and another one to 'mountain celery' (Greek: *oreoselinon*, *Athamanta macedonia* L.) (*MM* 3.64–65); and the *Alphabet of Galen* (chapter 297, p. 376 Everett) mentions both a wild (*erraticum*) and garden (*hortulanum*) variety of hyssop (*Hyssopus officinalis* L.). It is in gardens that the ancients grew their vegetables, herbs, common medicinal plants, and the flowers necessary for religious devotion. The kitchen garden was the basis of the ancient economy, as stressed by Jashemski: 'The hortus was primarily a kitchen garden, but I suspect that even so the ancient gardener tucked in a few flowers amid the herbs and vegetables, as does the Italian gardener today' (1981: 3).

6.2.2 The mountain forest and its plants

While the garden was seen as the epitome of cultivation in the ancient world, the mountain forest was the epitome of wilderness (see Amigues 1980 for an introduction).[3] One Latin word used to designate wild plants is '*silvestrus*', which literally means 'of the forest' (*silva*). In addition, plants that are qualified as 'mountain plants' (Greek: *oreios*; Latin: *montanus*) are usually wild plants, as in the case of the 'mountain celery' mentioned above or of a 'mountain' variety of southernwood mentioned by Pliny (*HN* 21.160). While the forest and the mountain may seem to be two completely different environments, it must be noted that the largest wild forests of the Mediterranean are located in mountain ranges. Thus, the poet Theocritus describes Mount Ida (of Troad) as 'much-wooded' (*poludendron*, *Idyll* 17.9). The northern forest the Romands had encountered in their conquests (in particular those of Germany) were also mountain forests.

Mountain forests tend to be cold, and ancient authors sometimes wondered why plants would seek such an environment. Pliny, in his usual anthropomorphizing style, for instance, writes: 'The pitch pine loves mountains and cold places ... Similarly the fir tree, which is sought after for shipbuilding, is found high up on mountains, as if it tried to escape the sea' (*HN* 16.40–42). Nicolaus attempts to explain why plants grow on mountains: 'Because mountains attract moisture and the clearness of the air assists the process, concoction is rapid. Therefore most plants grow on mountains' (*On Plants* 2.3.183, 825a32–34; translation: Drossaart Lulofs and Poortman 1989: 188).

As an environment, the mountain forest has attracted less scholarly interest than the garden (or other cultivated environments), even though to the ancients this was a most important environment. Theophrastus, for instance, devotes most of book three of the *Enquiry* to it. Ancient mountain forests provided the timber necessary for construction, and in particular for naval construction (see Amigues 1990–1991; Rival 1991; Allevato *et al.* 2009).[4] Forests also provided fuel (see Olson 1991) and resins, used in medicines and cosmetics (see André 1964).

During times of famine, they also gave food such as oak-acorns (see Galen, *Foodstuffs* 2.38, 6.620 Kühn). Finally, the most potent pharmacological plants used by the ancients were all collected on mountains. This, Theophrastus explains, is because of their particular ecology:

> And it would seem that [pharmacological plants] require a type of air that is cold but also pure, and again the right amount of nourishment. Certainly, it appears that most drugs grow on mountains, and particularly on the highest and greatest.
>
> (Theophrastus, *Causes of Plant Phaenomena* 6.13.5)

Dioscorides too stressed the quality of pharmacological plants growing on mountains: 'It also makes a big difference ... whether the places [from which you collect plants] are mountainous, high, windy, cold, and waterless. For the powers of such plants are stronger' (*MM*, preface 6). While these cool and breezy conditions were favourable to the growth of pharmacological plants, they were detrimental, according to Theophrastus, to the growth of sweet-smelling flowers (*HP* 6.8.6).

Thus, the forest, and in particular the mountain forest, was an important biotope from an economic point of view; one that provided 'free' produce to those who controlled it politically (see e.g. Xenophon, *Hellenica* 6.1.11). However, according to some scholars, the Mediterranean experienced severe deforestation during antiquity. We have already dealt with this topic in Chapter 1, indicating that definitive answers to the question require further multidisciplinary research. From our perspective, the case for deforestation seems weak for several reasons. First, many of the texts used by scholars postulating extensive deforestation are not particularly suited for drawing such conclusions. For instance, when Plato describes a wooded Attica of old in the *Critias* (111b–d), he is depicting a mythical environment. Second, this theory assumes that the ancients were unable to manage environments, and overexploited them. However, if this were the case, the ancients would not have developed complex rituals surrounding the gathering of wild plants (see Chapter 2 for more detail). Often, to the modern eye, these rituals seem irrational, but whatever their meaning, they must have acted as deterrents – or at least warnings – to the would-be plant gatherer. Third, the idea whereby many wild trees growing in forests (see Chapter 5) are male, hard, and difficult to deal with, may have contributed to the protection of the forest. The trees themselves resist overexploitation by displaying the characteristics the ancients would have attributed to some of their war enemies. Fourth, Theophrastus himself says that the regions that provided all the wood for naval construction were relatively small (*HP* 4.5.5). Wooded areas did not need to be large to provide the wood necessary for ancient building.

While we are not convinced by the case for catastrophic deforestation during antiquity, we do not deny that humans sometimes destroyed plant biotopes in antiquity; we will examine some examples of such destructions at the end of this chapter. Now, however, we turn to the topic of plant transplantation.

6.3 Ethnic vegetables and plant transplantation

The link between territory and plant was, as we already pointed out in Chapter 3, often stressed in the naming of plants after locations, as in the case of the 'Ethiopian cumin'. We have showed that the ancients used geographical epithets (descriptors) to distinguish between 'types' of plants. From the point of view of modern botany, these 'types' of plants are sometimes varieties within the same species; sometimes different species within the same genus; sometimes different genera within the same family; and sometimes they are completely unrelated.

Ancient botanical writers often gave lists of what we would call 'ethnic vegetables'. These lists were hierarchical, especially in the case of pharmacologically active plants. For instance, Theophrastus noted that the best black hellebore grew on Mount Helicon (see Map 1); while the best white hellebore came from four places: Mount Oeta (see Map 1), Pontus (see Map 1), Elea (see Map 4) and Malia (in Thessaly, see Map 1) (*HP* 9.10.3).[5] Elsewhere, he added that, while Mount Oeta and Mount Parnassus (see Map 1) are not particularly far from each other, the former produces excellent hellebore but the latter produces weak hellebore (*CP* 6.13.4). This habit of establishing hierarchies of pharmacological plants remained strong throughout antiquity. Thus, Dioscorides often gave such lists, as in the case of the camel hay: 'one [type] grows in Libya, another in Arabia, and another the [part of Arabia] called Nabataea, which is the best. The second best is the Arabian, which some call "Babylonian", but others "*teuchitis*". The Libyan is useless' (*MM* 1.17). Pliny too (who consulted the same sources as Dioscorides, see Chapter 2) established hierarchies of pharmacologically active plants, as in the following example:

> The most highly esteemed iris is found in Illyria, and even there not in the coastal districts, but in the woody parts near the Drinon and around Narona [see Map 3]. Next after it comes the Macedonian iris.... Third in estimation comes the African iris.... The Pisidian variety, too, is by no means despised.
>
> (Pliny, *Natural History* 21.40–42)

While in medicine, it might have been important to establish hierarchies, in other cases, these are more a question of taste. Hierarchies were also established in the case of vines, and the wines they produced (see e.g. Pliny, *HN* 14.21–53).

Initially, geographical epithets reflected the place of origin of a plant – or a place where a particularly good strand of a plant grew. Matters were, however, much more complicated, as the Greeks and Romans constantly sought to transplant plants, with varied levels of success. Our information on this topic is abundant and comes from various types of sources. First, the movement of plants from their indigenous environment to new ones can be uncovered through archaeological remains. For instance, the carbonised remains of a lemon tree (*Citrus limon* (L.) Burm f.) discovered by Wilhelmina Jashemski at the villa Poppaea at Oplontis (near Pompeii) shows that the lemon was growing in the area (Jashemski 1999: 38).[6] Sadori *et al.* (2009) have researched the introduction and diffusion of the peach (*Prunus persica* L.) in Italy, using both iconography and archaeological remains. They concluded that the peach was introduced in Italy in the first half of the first century CE. Bakels and Jacomet (2003) have traced the movement of various plants (walnut, apple, pear, plum, cherry, garlic) from Rome to the western parts of the Roman Empire.

Second, documentary papyri (that is, papyri that do not contain literary texts) can offer invaluable information on plant transplantation. Thus, the Zeno-archive, a significant collection of papyrological documents relating to Zeno of Caunus (the secretary of Apollonius, an official under the rule of Ptolemy II and Ptolemy III in Hellenistic Egypt) documents several plant transplants. For instance, a letter from a certain Euempolus to Zeno (250 BCE) mentions a garlic of Tlos, a city in Lycia (see Map 1), whence this variety had most probably been imported (P. Cairo Zen. 3.59299 = PSI 4.433 = Trismegistos 943; see Crawford 1973; Thompson 1984: 367). This papyrus also mentions a garlic 'from the oasis', probably referring to an Egyptian locality. Finally, there are numerous ancient literary accounts of plant transplantations. These are often very detailed, sometimes suspiciously so, in particular in the writings of Pliny. Here is the encyclopaedist's description of the plane tree's peregrinations:

> But who would not justifiably be surprised at the fact that a tree has been sought from abroad merely for the sake of shade? This tree is the plane, which was first brought into the Ionian Sea, as far as the island of Diomedes [San Domino in the Tremiti islands in the Adriatic; see Map 4], to be planted on the tomb of Diomedes; thence it crossed to Sicily and was among the first trees donated to Italy. And now it has travelled as far as Belgic Gaul [see Map 3], and it occupies soil subject to tribute, so that people even pay a tax for shade! Dionysius, the old tyrant of Sicily, imported the plane to the city of Rhegium [see Map 4] as a marvel for his own house, at the place where a gymnasium was built afterwards; and it is found in the authorities that these trees could not grow to full size, and that in Italy there were no other except the Hispania variety.
>
> (Pliny the Elder, *Natural History* 12.6–7[7])

Note here the blending of botanical information with political vocabulary; the mention of the tribute and the tax. Note in particular the mention of Dionysius, that is, Dionysius I of Syracuse (fifth–fourth century BCE). Indeed, Pliny often links plant transplantations to political figures. Thus, Lucullus (Lucius Licinus Lucullus, first century BCE) allegedly transplanted the cherry tree from Pontus into Italy; by the time of Pliny that tree had 'crossed the ocean and got as far as Britain'[8] (Pliny, *HN* 15.102); the consul Sextus Papinius introduced the jujube tree and the tuber apple from Africa into Italy in 23 CE (*HN* 15.47); and Lucius Vitellius imported the pistachio from Syria to Alba (*HN* 15.91; see Map 4 for the location of Alba).

Some transplants had been so successful that the ethnic epithets attached to plants names had become all but meaningless. For instance, by the time of Pliny, the quince, named *cydonea* after its place of origin, Cydonia in Crete, grew beautifully in Naples (*HN* 15.37); and the damson, named *damascena* in Latin, after Damascus in Syria, also grew in Italy, though with a larger stone and less flesh (*HN* 15.43). Pliny also noted that Cato called the female cypress, which originated in Crete, the Taranto cypress 'no doubt because that place was where it was first imported' (*HN* 16.142; see Map 4). Columella, for his part, boasted of how Italy 'had learnt to carry the fruits of almost the entire world, as its inhabitants zealously applied themselves to the task' (*RR* 3.8.5).

Transplanting was by no means easy, and it could take a long time to achieve success. In the case of transplanted cereals, according to Theophrastus, one had to allow three years for the imported crop to adapt to its new environment:

> Foreign (*xenika*) grains change into native ones (*epichōriois*) on average within three years.... For that reason they say that one must be careful not to mix foreign grains with native ones, unless they come from similar [localities], for they are at variance with the region.
>
> (Theophrastus, *Enquiry into Plants* 8.8.1)

Theophrastus makes the same point in *Causes of Plant Phaenomena*, there comparing plants to animals, and stating that the change from foreign to native occurs after three generations (*trigonēsanta*, *CP* 2.13.3). He also adds that the process is essentially the same in grains as in trees, even though trees live much longer than grains:

> The difference [between the adaptation of tree and grain] consists in the fact that the tree, once planted, always remains the same [individual], while the grain changes from one individual to another. That said, the process is the same: for it [the region] takes away from the nature of the plant, by equally mastering both [the grain and the tree].
>
> (Theophrastus, *Causes of Plant Phaenomena* 2.13.4–5)

Aristotle had made a similar point in his *Generation of Animals*, when discussing the hybridisation of animals. He argued that with the succession of generations, the hybrid starts taking on the female form, 'just as in the case of foreign (*xenika*) seeds that take after the region (*chōran*) that provides the material, that is, the body, for the seeds' (*GA* 2.4, 738b34–36). It is the nature of the soil that could sometimes lead to a transplant failure, according to the author of the

Hippocratic treatise *Diseases* IV (the author who also wrote *Generation* and *Nature of the Child*). This writer notes that numerous attempts to grow silphium in Ionia and the Peloponnesus failed because it could not find the humour it needed for its growth.

> Nevertheless it was not possible, despite numerous attempts, to grow silphium either in Ionia or in the Peloponnese. In Libya, on the other hand, it grows spontaneously. This is because neither in Ionia nor in the Peloponnesus is there a humour (*ikma*) such as is able to nourish this plant.
>
> (*Diseases* 4.34, Loeb 10.104 Potter = 7.546 Littré)

The author goes on to say that even transplantation over extremely small distances can fail because the plant will not find the right 'humour' to sustain its growth in its new locale.

Some transplants were doomed to failure: for instance the attempt by Harpalus, a friend and courtier of Alexander the Great, to grow ivy in Syria and Babylonia (Theophrastus, *HP* 4.4.1; see also Plutarch, *Table Talk* 3.2.1, 648c; Pliny, *HN* 16.144). Other transplants transformed plants. For instance, Pliny tells us that the Egyptian plant lotus 'has become familiarised (*familiarem*) with Italy, but has been changed by the [nature of] the soil' (*HN* 13.104). One of the most commonly repeated examples of half-successful transplantation was that of the date palm. The palm grew barren in the Aegean world, which defied the object of transplantation because trees were imported for their useful fruits. Theophrastus notes that if one were to transplant a barren palm tree back in its original region, it would bear fruit again:

> It is also probable that, if someone were to plant the palm tree that grows here [literally: our palm tree] in Babylon, it would bear fruits and become exactly like that growing there. And it would be the same case with other countries that bear fruits that are congenial to that region. For location is more important than cultivation and care. A sign of this is that plants transplanted thence become unfruitful and in some cases do not even grow.
>
> (Theophrastus, *Enquiry into Plants* 2.2.8)

Similar comments are made by Pliny (see p. 172), Varro (*RR* 2.1.27) and other agronomists. Greeks and Romans considered the palm to be extremely peculiar because it enjoyed a briny soil, when salt was known to destroy many plants. They commented on the Babylonian habit of watering the tree with briny water and 'manuring' the ground in which it grows with lumps of salt (see e.g. Theophrastus, *CP* 3.17.1–4; Palladius, *OA* 11.12.2).

Ancient stories of plant transplantations – successful or otherwise – could be multiplied. We leave this topic with a passage from Pliny, where the encyclopaedist (as per usual) gives anthropomorphising accounts of plant behaviours:

> We have shown that not all plants grow in all places, and that when they are transferred they do not always live. This is due in some cases to pride (*fastidio*), in others to obstinacy (*contumacia*), more often to the weakness of the species that are transplanted, in some cases to the unfavourable climate, in others to the incompatibility of the soil. The balm tree despises (*fastidit*) the very idea of growing elsewhere; and a fruit tree born in Assyria will not bear elsewhere; and similarly the palm tree will not grow everywhere, or even if it does grow, it will not bear, and when it has made a promise and a show of bearing fruit, it does as if it had given birth against its will (*invita*).... What is most surprising is that, although the trees can sometimes be persuaded (*exorari*) to live and to migrate, and on occasions will be granted by the soil the request to feed foreigners (*alienas*) and nourish immigrants (*advenas*), the climate is never bent.
>
> (Pliny, *Natural History* 16.134–136)

Pliny's vocabulary to describe transplanted plants is the one he would have used to describe human slaves. For him plants are like vanquished foreigners who refused to be taken away from their native country in servitude. They have human-like feelings, and in particular pride. Yet, plants' will can usually be bent, whereas a hostile environment will remain so.

So far in this chapter (and in general in this book), we have discussed plants that grow on land. However, the ancients were also interested in aquatic plants. In the following section, we turn to these, and in particular to marine plants.

6.4 Aquatic plants

To Theophrastus, the *diaphora* between terrestrial and aquatic plants was a significant one, which he introduced early on in the *Enquiry*, one that corresponded to the *diaphora* found in animals (*HP* 1.4.2). Further to this basic classification, the philosopher distinguished between various types of aquatic plants:

> For there are some plants that cannot live except in the wet. And they are distinguished one from another by the type of wet [they live in], so that some grow in ponds, others in marshy lakes, some in rivers, and others even in the sea – the smallest of those in our sea [the Mediterranean], and the bigger ones in the Erythrean Sea [Red Sea]. Others still are plants, as it were, of extremely wet places and marshes, such as the willow and the plane tree.
>
> (Theophrastus, *Enquiry into Plants* 1.4.2)

A few lines later, Theophrastus even mentions amphibious (*amphibia*) plants (*HP* 1.4.3). Theophrastus concentrates most of his discussion on aquatic plants at the end of book four of the *Enquiry*, where he discusses marine plants growing in the Mediterranean (chapter 6); marine plants growing beyond the Pillars of Heracles, and to the east of the Greek world, including mangrove plants (chapter 7); aquatic plants growing in Egypt (including papyrus, chapter 8); the water chestnut (chapter 9); aquatic plants growing in Greece, and in particular in Orchomenus Lake (chapter 10); and reeds and rushes (chapters 11 and 12). Here we shall focus on marine plants, as they were a puzzle to ancient authors because they grew in salt water. Salt water was seen as particularly inimical to the growth of terrestrial plants (with the exception of the palm tree, which sought briny soil). Theophrastus solved the conundrum in the following manner:

> The briny is not nutritious and, as it were, non-productive. The proof is that nothing – so to say – will grow in such regions [i.e. regions were the soil is salty, deserts]. For the salt destroys the plant's powers and thus prevents it from forming. Now it seems logical that that which causes destruction in other things will not generate itself. But even the plants that grow in the sea grow and are formed by a certain type of sweetness and by other flavours (and the same goes for fish and other animals of the sea).
>
> (Theophrastus, *Causes of Plant Phaenomena* 6.10.1–2)

In other words, although 'salty' is the prevalent 'taste' in sea water, it also contains other tastes. It is those that allow plants to grow in the sea. Plutarch in his *Natural Questions* also attempted to answer the question 'why does sea water not nourish trees?' He mostly focused on the fact that sea water kills terrestrial plants, citing various ancient philosophers, but ended his paragraph with the following note:

> And the shrubs and plants which they say bloom in the Erythrean Sea, they bear no fruit, and they are nourished by the rivers that throw much mud [into the sea]. Hence they grow not far from land, but near it.
>
> (Plutarch, *Natural Questions* 1.1, 911e–f)

Plutarch, then, believes that marine plants are fed by the water of rivers and the earth contained therein. In his opinion, algae cannot grow far from shore; this, however, is not the case.

Finally, the medical author Aetius, in a unique and particularly interesting passage, comments on the saltiness of the Dead Sea (see Map 3):

> For it appears that neither animal nor plant is born in that water [the Dead Sea]. But while the rivers that empty themselves into it have numerous large fish, in particular the holy stream nearby, which they call Jordan, no fish swims beyond the mouth of the river. If a hunter were to throw one [fish] in that lake, he would see it being killed quickly. For that reason they call it the 'Dead Lake', for it is in fact a

> lagoon. Thus extremely salty [environments] are hostile to all plants and animals, and by nature they are dry and drying.
>
> (Aetius, *Medical Collection* 1, preface, CMG 8.1, p. 27 Olivieri)

The Dead Sea is indeed very hostile to life, although it may produce some microscopic bacteria and algae, which Aetius would not have been able to observe.

Theophrastus gives us some good descriptions of marine plants. Most of the algae he mentions are named after terrestrial plants: e.g. the 'sea fir', 'sea fig', 'sea oak' 'sea vine' and 'sea palm' (*HP* 4.6.2). These names have remained until today as vernacular names of algae. For example, the brown seaweed *Halidrys siliquosa* (L.) Lyngb. is known as sea oak; and so is the red alga *Phycodrys rubens* (L.) Batt., whose leaf looks remarkably like that of an oak tree. The species described by Theophrastus are all large, conspicuous species which would be apparent growing on the shore or have been cast up from the sublittoral zone (i.e. below low water mark). However, on one occasion Theophrastus describes drift specimens of seaweeds, that is, specimens which have become detacted. Drift specimens are particularly common on the shore following rough weather and storms:

> In the Outer Sea [i.e. Atlantic Ocean], around the pillars of Heracles [i.e. Straights of Gibraltar], there is a type [of *phukos*] of amazing size, as they say, and larger than a palm in breadth. This is carried into the Inner Sea [i.e. the Mediterranean] with the current from the Outer Sea, and they call it 'leek'. And in that sea, in some places, it is higher than a man's waist. It is said to be annual, to grow towards the end of spring, and to be in its prime during the summer, wither in the autumn, and die in the winter, when it is thrown out [to shore].
>
> (Theophrastus, *Enquiry into Plants* 4.6.4)

This description corresponds quite closely to that of a kelp, perhaps *Laminaria saccharina* (L.) Lamour. or *Saccorhiza polyschides* (Lightf.) Batt. (see Amigues 1989: 240 for the latter identification). The holdfast and stipe (i.e. the root-like and stalk-like organisms of a seaweed) of most kelps persist for several years, a new blade being produced each spring, flourishing during the summer, and dying off in the autumn or winter; Theophrastus' observations are therefore remarkably accurate. Eventually the stipes and holdfasts become detached (usually during winter storms) and are 'thrown up on shore' as stated by Theophrastus.

Importantly, all the marine species Theophrastus describes are plants. He states that: 'sponges, the usual types and those that are called "unwashable", and organisms of the same type are different in nature' (*HP* 4.6.10). There was, indeed, much confusion as to the classification of marine organisms in the ancient world (see Chapter 3): were they plants or animals? Pliny's discussion of marine plants (which is otherwise mostly based on Theophrastus' account) includes this following account, which appears to describe animals:

> Juba [II of Mauretania] reports that around the islands of the Cave-Dwellers, a bush grows deep in the sea, called 'hair of Isis'. It is similar to coral and has no leaves. When it is cut off, it changes its colour to black and hardens; and when it falls, it shatters. The same happens to another [bush] which is called 'eyelid of the Graces', which is a powerful aphrodisiac. He reports that women make bracelets and necklaces out of it. He says that this bush can feel when it is taken and hardens like a horn, thus blunting the edge of the knife, but that, if it is cut unwittingly, it changes into stone.
>
> (Pliny, *Natural History* 13.142)

It is unclear what exactly Juba, the learned king of Mauretania, had in mind when describing these organisms, but they appear to be species of coral – animals rather than plants. However, the confusion regarding the nature of corals was common in the ancient world, and it persists in 'popular' thought. Indeed, one still speaks of 'coral gardens' and 'coral gathering'.

The ancients had no international rules of conservation of species, which are applied to many marine species, but they were aware of human impact on biotopes. We now turn to some examples of such impact discussed in ancient botanical sources.

6.5 Plant conservation issues in antiquity

We have already discussed the question of possible extensive deforestation in antiquity. In our opinion, the evidence is too slim to substantiate the claim whereby the Mediterranean witnessed widespread deforestation. One of the sources that is regularly quoted in discussions of deforestation and destruction of Mediterranean maquis is a fragment from the comedy *The Goats* by Eupolis (see e.g. Hughes 2006: 240; Sands 2013: 16).[9] This fragment is a passage where the goats (which constitute the chorus of the play) boast about their destructive power:

> We feed on trees of all sorts: fir, prickly oak, strawberry tree,
> Biting off the tender shoots, and also their branches,
> tree medick, sweet-smelling sage, many-leaved bindweed,
> Wild olive, mastic, manna ash, white poplar, holm oak, oak, ivy, heather,
> Willow, buckthorn, mullein, asphodel, Valonia oak, rock rose, thyme
> Savory.
>
> (Eupolis fr. 13 Kassel and Austin = Plutarch, *Table Talk* 4.1.3, 662d)

It is important to stress that this is a comedy passage – it is meant to make audiences laugh. Long enumerations have the power to provoke hilarity, especially when they are filled with puns and other jokes. One must not take this passage literally.

The ancients were perfectly aware that goats could be destructive (see Chapter 5 for more detail), and they knew it was not in their interest to let them overgraze. We suspect overgrazing was extremely rarely allowed to occur in antiquity. It is, however, blamed for the disappearance of silphium by Pliny:

> It has not been found in this land [Cyrene] for many years, because the tax-farmers who rent the pasture-land destroy it by grazing sheep on it, believing that they would make more profit this way. There has only been one stem found in living memory; it was sent to the Emperor Nero.
>
> (Pliny, *Natural History* 19.39 (see Totelin 2015, with the same translation))

We find this passage of Pliny rather puzzling. Silphium was an expensive produce and it contributed greatly to the economy of Cyrene, as indicated by its presence on numerous coins (see Chapter 4). It is highly improbably that tax-farmers considered silphium to be less valuable than grazing sheep. If silphium did indeed become extinct – opinions are divided on the topic – the reason may have to be found elsewhere. Whether the cause of that possible extinction is overexploitation or some ecological cause (a disease, desertification) is impossible to determine (on the possible extinction of silphium, see: Andrews 1941; Roques 1984, 1993; Parejko 2003).

We are not arguing that humans in antiquity never had a detrimental impact on plant species, but rather that one has to be extremely careful when analysing written sources. One always has to take into account the conventions of genres and the biases of authors. There are in fact some very clear indications that humans sometimes altered their environment.[10] Thus, Theophrastus notes several man-made climatic changes in his *Causes of Plant Phaenomena*:

> In the region around Larissa in Thessaly [see Map 1], formerly there was much standing water and the plain was marshy, the air was thick and the region was warmer. But now that the water has been drained and prevented from accumulating, the region has become colder, and it freezes more often. The proof, as they [the inhabitants of Larissa?] say, is that there were once beautiful and large olive trees in the city itself and elsewhere, but now there are none. And the vines never used to freeze, but now they do so often…
>
> In Philippi [see Map 1], formerly freezing occurred more often, but now that most of the water has disappeared underground and dried out, and the whole region has become cultivated, there is much less. And yet the air is thinner on account of these two reasons – the drying out of the water and the cultivation of the region. For uncultivated land is colder and has a thicker air because it is wooded and the sunlight cannot reach it, and the winds cannot aerate it, and at the same time, [uncultivated

> land] has much stagnating and accumulating water. And this is what happened at Crenides [see Map 1] when the Thracians inhabited that region. For the entire plain was covered with trees and water.
>
> (Theophrastus, *Causes of Plant Phaenomena* 5.14.2–6)

These passages clearly describe environmental changes brought on by drainage of stagnating water for the purpose of cultivation, and in particular of cereal cultivation. This certainly had a deep impact on the trees of the region. Note, however, that the changes are extremely localised. Note, too, that Pliny did not have many more examples of such environmental changes to add to those of Theophrastus when he wrote some 400 years later (*HN* 17.30). Such radical – but localised – environmental changes appear to have been relatively rare in antiquity.

6.6 Conclusions

Humans have always sought to control their environment in general, and more particularly its plants. Greek and Roman authors talk of plants as if they were war enemies to be controlled, barbarians who refuse to be transplanted and to be productive for their masters, barbarians who reject the benefits of culture/cultivation. Ancient botanical authors certainly believed that cultivation was a force for good. Their statements, however, were not unmitigated. In ancient polarised thought, the cultivated was characterised as female, and the wild as male. One could not exist without the other; and wild plants, when submitted to cultivation, became effeminate. If forests could be seen as threatening, inhabited as they were with hardened trees and pharmacological plants, they were also extremely useful, perhaps more useful than the ornamental gardens that had become so fashionable in the first centuries of the Common Era.

One notion that is particularly prominent in ancient 'ecological' thought, is the notion of beauty. When a plant is in its proper environment, its proper home (*oikos*), it grows most beautiful and most useful. The ancients may have attempted to bend their environment to their needs, but ultimately, they understood that nature was stronger and more powerful than the mightiest of kings.

Notes

1 On the environments not discussed here, see e.g. White 1970: 173–271 (in particular for cereals and orchards); Amouretti 1986 (on cereals and olive); Sallares 1991: 294–389 (agricultural products in general); Amouretti and Brun 1993 (on vine and olive); Foxhall 2007 (olive). On food more generally, see Garnsey 1999; Dalby 2003.

2 For a general introduction to ancient gardens, see Carroll-Spillecke 1992 and 2003. On Greek gardens, see Thompson and Griswold 1963; Caroll-Spillecke 1989 (there are also numerous articles on Greek gardens, but space prevents us from offering a full bibliography). On Hellenistic gardens, see in particular Nielsen 2001. On Roman gardens, see e.g. Jashemski 1979; MacDougall and Jashemski 1981; Grimal 1984; Cima and La Rocca 1998; Farrar 1998; Ciarallo 2001; Bowe 2004; Frass 2006; von

Stackelberg 2009. On Byzantine gardens, see e.g. Littlewood *et al.* 2002. We have only included here monographs and collections of essays on the Roman garden (and our list is not exhaustive). There are beside these many articles on the topic. For a collection of Latin texts on gardens and gardening translated into English, see Henderson 2004.

3 See also Delano Smith 1996 on the question of what constituted 'wilderness' to the Romans. See Thommen (2012: 85–89) on Roman attitudes to the forest.

4 For the use of wood in other types of construction, see Adam 1994 as an introduction.

5 Theophrastus, at *HP* 9.15.1–5, also offers a list of the best drug-producing places within the Greek world (Mount Pelion of Thessaly; the Telethrion in Euboea; Mount Parnassus; Arcadia, and Laconia; see Map 1) and outside the Greek world (Etruria, Latium, Egypt, Ethiopia, a territory much larger than modern Ethiopia, Scythia, India, and Thrace).

6 It is sometimes thought that the Greeks and Roman did not know the lemon, and only grew citrons (another type of citrus fruit, *Citrus medica* L.), but Jashemski asserts that the archaeobotanic evidence as well as the archaeological evidence (in the form of mosaics that appear to depict lemons rather than citrons) confirm the presence of the citron in the Roman world by the first century CE. See contra Zohary *et al.* 2013: 7 (the lemon was introduced by the Arabs). On the diffusion of citrus fruits in the Mediterranean, see also Coubray *et al.* 2010; Pagnoux *et al.* 2013.

7 On the plane tree and its links to the orient, see Stubbings 1946.

8 There were 'cherries' in Italy well before the time of Lucullus, but these might all have been Cornelian cherries (*Cornus mas* L.) rather than our sweet cherries (*Prunus avium* L.). See Isidore, *Etymologies* 17.7.16 on this topic. On cherries in Britain in an archaeological context, see van der Veen *et al.* 2008, especially p. 27.

9 Wertime 1983 argues that the goat was not as damaging to the ancient forest as were pyrotechnologic industries.

10 See Rackham 1996: 33–35 for interesting comments on the question 'Did the ancient Greeks have an attitude to ecology?' See also Amigues 1991 for different examples than those developed here.

CONCLUSIONS

Useful and wonderful plants

Quintilian (Marcus Fabius Quintilianus) was a first-century CE Roman orator. Under his name – but almost certainly not authentic – is an oration entitled *The Poor Man's Bees* (*Apes pauperis*). The argument of the oration goes as follows:

> There may be legal suit for damage caused by wrongdoing: a poor man and a rich man were neighbours in the country by their adjacent gardens (*hortulis*). The rich had flowers in his garden; the poor had bees. The rich man complained that his flowers were damaged by the bees of the poor man. He requested for them to be relocated. As the poor man did not transfer them, the rich man sprinkled his flowers with poison. All the bees of the poor man perished. The rich man is accused of damage caused by wrongdoing.
>
> (pseudo-Quintilian, *Orations* 13.1)

The Greeks and Romans did not understand pollination by insects, and enshrined this misunderstanding in law. Today, with the accelerating extinction of bees, the rich man would probably thank his poor neighbour for the service his hives were providing. In antiquity, by contrast, the bees were seen as 'stealing' nectar to produce honey that would not benefit the rich man.

The Swedish botanist Carl Linnaeus (1707–1778) famously put plant sexuality at the centre of his understanding of plants (he was not the first, but he certainly was the most successful at doing this). People who have not studied biology may not know what carpels and stamens are, but they will all be familiar with the euphemism 'the birds and the bees', where the bees refer to plant pollination and sexuality. In modern botany, it is understood that plants form their own 'kingdom' with its own rules; one that uses animals such as bees for its own survival.

In antiquity, Aristotle argued that the *telos* of plants, their purpose, was the production of seed. In some ways, he too, put generation at the centre of his understanding of plants. However, beside the fact that ancient plant generation is very different from plant sexuality, Aristotle considered plants to be at the bottom of the chain of nature. Plants were useful to animals, and in particular to humans: their seeds would serve as food and medicine for these higher beings; their fibres would serve as clothing; their wood as building material.

In this book, we have argued that 'utility' is one of the central notions of ancient botanical thought. It is so central that it makes no sense to divide ancient botany into 'pure' and 'applied'. In a way, all ancient botany is applied botany. Our purpose here has been to place botanical thought in its social context. In order to do so, we have explored a large variety of texts, ranging from philosophical ones to poetical satire and comedy – there is no doubt that we could have discussed many more. All the authors we discussed displayed an excellent knowledge of plants, and often an awareness of difficult philosophical questions, such as whether plants are sentient or not. Beyond these preserved authors, there were many informants, people who often were illiterate: farmers, root-cutters, wreath-makers, sponge-divers and woodcutters. These were the people who made a living from plants, or perhaps struggled to make a living from plants. They had to contend with rich people who had all the resources necessary to buy extensive estates, transplant plants, fell large wooded areas, and create decorative – but unproductive – gardens.

Poor and rich alike, however, also had to contend with the trickiness of plants and nature. For plants were tricky: tricky to define, to name, to describe, to classify, to bend to cultivation, to transplant, and to use as medicine without causing poisoning. Cultivated plants often behaved like disobedient children or slaves: eating too much and wasting their master's energy and time. Wild plants, for their part, behaved like war enemies: proud and untameable. Plants, at the bottom of the great chain of being, certainly caused trouble for humans at the top of the scale.

Ancient botanical authors did complain about the difficulties posed by plants, but they also reacted to these in a more positive manner: with wonder and amazement. Another important notion in ancient botanical thought is that of beauty. Beautiful plants can grow on the coldest mountain, in the saltiest desert, and in the deepest sea; they can outlive the longest-lived animal; their shoots can grow new roots; they can display all the colours of the rainbow.

We started writing this book with a love of plants and a love of ancient technical texts. We found the wonderment of ancient authors contagious, and our passion grew with writing. We have approached the topic from numerous angles, and at each new turn we found new pieces of fascination. We hope to have shown that it is more fruitful to consider ancient botany on its own terms rather than to compare it to modern (post-Linnaean) botany. We have pointed to many areas where research remains to be done. Like pseudo-Quintilian's bees, we will carry on stealing the nectar of ancient flowers. We hope you will do too.

BIBLIOGRAPHY

Abbe, E. (1965). *The Plants of Virgil's Georgics: Commentary and Woodcuts*. Ithaca, NY: Cornell University Press.

Adam, J.-P. (1994). *Roman Building. Materials and Techniques. Translated by Anthony Matthews*. London: B. T. Batsford.

Adams, F. (1847). *The Seven Books of Paulus Aegineta*. London: The Sydenham Society.

Adler, A. (1928–1935). *Suidae lexicon*. 4 vols. Leipzig: Teubner.

Ainsworth, G. C. (1976). *Introduction to the History of Mycology*. Cambridge: Cambridge University Press.

Aliotta, G., Piomelli, D., Pollio, A. and Touwaide, A. (2003). *Le piante medicinali del Corpus Hippocraticum*. Naples: Guerini e associati.

Allevato, E., Russo Ermolli, E. and Di Pasquale, G. (2009). 'Woodland exploitation and Roman shipbuilding: preliminary data from the shipwreck Napoli C (Naples, Italy)'. *Méditerranée*, 112(1), 33–42.

Amigues, S. (1980). 'Quelques aspects de la forêt dans la littérature grecque antique'. *Revue forestière française*, 32(2), 211–223. Reprinted in Amigues 2002, 225–242.

Amigues, S. (1984). 'Phytonymes grecs et morphologie végétale'. *Journal des Savants*, 151–173. Reprinted in Amigues 2002, 291–310.

Amigues, S. (1986). 'Sur le perséa d'Egypte'. In *Hommages à François Daumas* (pp. 25–31). Montpellier: Université Paul Valéry. Institut d'Egyptologie. Reprinted in Amigues 2002, 141–148.

Amigues, S. (1988a). *Théophraste. Recherches sur les plantes. Tome I. Livres I-II. Texte établi et traduit par S. Amigues*. Paris: Les Belles Lettres.

Amigues, S. (1988b). 'Le crocus et le safran sur une fresque de Théra'. *Revue Archéologique*, 1988(2), 227–242. Reprinted in Amigues 2002, 95–112.

Amigues, S. (1989). *Théophraste. Recherches sur les plantes. Tome II. Livres III–IV. Texte établi et traduit par S. Amigues*. Paris: Les Belles Lettres.

Amigues, S. (1990–1991). 'Termes techniques de construction navale dans Théophraste, HP V, 7, 1–3'. *Revue Archéologique*, 85–96. Reprinted in Amigues 2002, 79–94.

Amigues, S. (1991), 'Le témoignage de l'antiquité classique sur des espèces en régression'. *Revue Forestière Française*, 43(1), 47–58. Reprinted in Amigues 2002, 263–276.

Amigues, S. (1992). 'Le rôle de la métaphore dans la formation des noms grecs de plantes'. In S. Gély (Ed.), *Sens et pouvoirs de la nomination dans les cultures hellénique et romaine. Tome II. Le nom et la métamorphose* (pp. 295–308). Montpellier: Université Paul Valéry. Reprinted in Amigues 2002, 279–290.

Amigues, S. (1993). *Théophraste. Recherches sur les plantes. Tome III. Livres V–VI. Texte établi et traduit par S. Amigues*. Paris: Les Belles Lettres.

Amigues, S. (1994). 'A l'origine de la botanique, les *Recherches sur les plantes* de Théophraste'. *Bulletin de l'Académie des Sciences et Lettres de Montpellier*, 25, 167–178. Reprinted in Amigues 2002, 3–10.

Amigues, S. (1995a). 'Végétation et cultures du proche-orient dans l'*Anabase*'. *Pallas*, 43, 61–78. Reprinted in Amigues 2002, 127–140.

Amigues, S. (1995b). 'Des plantes nommées *moly*'. *Journal des Savants*, 3–29. Reprinted in Amigues 2002, 429–452.

Amigues, S. (1995c), 'Les plantes d'Egypte vues par les naturalistes grecs'. In J. Leclant (Ed.), *Entre Egypte et Grèce: actes du Colloque du 6–9 octobre 1994* (pp. 51–67). Paris: Académie des Inscriptions et Belles-Lettres. Reprinted in Amigues 2002, 247–261.

Amigues, S. (1996a). 'De la botanique à la poésie dans les *Idylles* de Théocrite'. *Revue des Etudes Grecques*, 109(2), 467–488. Reprinted in Amigues 2002, 363–378.

Amigues, S. (1996b). 'L'expédition d'Anaxicrate en Arabie occidentale'. *Topoi Orient-Occident*, 6, 671–677. Reprinted in Amigues 2002, 57–62.

Amigues, S. (1996c). 'Les "fiancées du soleil"'. In A. Moreau and J.-C. Turpin, *Les Astres: actes du Colloque international de Montpellier, 23–25 mars 1995* (pp. 101–119). Montpellier: Université Paul Valéry. Reprinted in Amigues 2002, 379–393.

Amigues, S. (1998). 'Problèmes de composition et de classification dans l'*Historia plantarum* de Théophraste'. In J. M. van Ophuijsen and M. van Raalte (Eds), *Theophrastus: Reappraising the Sources* (pp. 191–202). New Brunswick and London: Transaction Publishers. Reprinted in Amigues 2002, 45–54.

Amigues, S. (1999a). 'Les traités botaniques de Théophraste'. In G. Wöhrle (Ed.), *Geschichte der Mathematik und der Naturwissenschaften in der Antike. Band I: Biologie* (pp. 124–154). Stuttgart: F. Steiner. Reprinted in Amigues 2002, 11–44.

Amigues, S. (1999b). 'Note sur l'"éllébore blanc" de Dioscorides, IV, 148'. *Revue de Philologie*, 73(1), 7–13. Reprinted in Amigues 2002, 177–184.

Amigues, S. (2000). 'Ambiguités et vicissitudes des noms de plantes de Théophraste à Linné'. *Bulletin de l'Académie des Sciences et Lettres de Montpellier*, 31, 53–61. Reprinted in Amigues 2002, 351–362.

Amigues, S. (2002). *Etudes de botanique antique*. Paris: Diffusion De Boccard.

Amigues, S. (2003). *Théophraste. Recherches sur les plantes. Tome IV. Livres VII et VIII. Texte établi et traduit par S. Amigues*. Paris: Les Belles Lettres.

Amigues, S. (2004a). 'Le silphium, état de la question'. *Journal des Savants*, 191–226.

Amigues, S. (2004b). 'Du jujubier des Lotophages à l'arbre sacré du Temple d'Or'. In E. Ciurtin (Ed.), *Du corps humain, au carrefour de plusieurs savoirs en Inde. Mélanges offerts à Arion Rosu par ses collègues et ses amis à l'occasion de son 80e anniversaire* (pp. 51–68). Paris: De Boccard.

Amigues, S. (2006). *Théophraste. Recherches sur les plantes. Tome V. Livre IX. Texte établi et traduit par S. Amigues*. Paris: Les Belles Lettres.

Amigues, S. (2007). 'L'"arbre à styrax": nouvelle interprétation d'un monnayage de Selgè'. *Comptes-rendus des séances de l'Académie des Inscriptions et Belles-Lettres*, 151(1), 285–293.

Amigues, S. (2010). *Théophraste Recherches sur les plantes. À l'origine de la botanique. Préface de Paul Bernard*. Paris: Belin.

Amigues, S. (2012). *Théophratse. Les causes des phénomènes végétaux. Tome I. Livres I et II. Texte établi et traduit par S. Amigues*. Paris: Les Belles Lettres.

BIBLIOGRAPHY

Amouretti, M.-C. (1986). *Le pain et l'huile dans la Grèce antique. De l'araire au moulin.* Paris Annales littéraires de l'Univeristé de Besançon.

Amouretti, M.-C. and Brun, J.-P. (1993). *La Production du vin et et de l'huile en Méditerranée.* Athens and Paris: De Boccard.

Andersen, Ø. (1976). 'Aristotle on sense-perception in plants'. *Symbolae Osloenses*, 51(1), 81–86.

Anderson, F. J. (1977). *An Illustrated History of the Herbals.* New York: Columbia University Press.

Andrews, A. C. (1941). 'The silphium of the ancients: a lesson in crop control'. *Isis*, 33, 232–236.

André, J. (1954). 'Les noms latins de l'hellébore'. *Revue des Etudes Latines*, 32, 174–182.

André, J. (1955). 'Pline l'ancien botaniste'. *Revue des Etudes Latines*, 33, 297–318.

André, J. (1958). *Pline l'Ancien. Histoire naturelle. Livre XIV. Texte établi, traduit et commenté par J. André.* Paris: Les Belles Lettres.

André, J. (1960). *Pline l'Ancien. Histoire naturelle. Livre XV. Texte établi, traduit et commenté par J. André.* Paris: Les Belles Lettres.

André, J. (1962). *Pline l'ancien. Histoire naturelle. Livre XVI. Texte établi, traduit et commenté par J. André.* Paris: Les Belles Lettres.

André, J. (1964a). *Pline l'Ancien. Histoire naturelle. Livre XVII. Texte établi, traduit et commenté par J. André.* Paris: Les Belles Lettres.

André, J. (1964b). *Pline l'Ancien. Histoire naturelle. Livre XIX. Texte établi et traduit par J. André.* Paris: Les Belles Lettres.

André, J. (1964c). 'La résine et la poix dans l'antiquité. Technique et terminologie'. *L'Antiquité classique*, 33(1), 86–97.

André, J. (1965). *Pline l'Ancien. Histoire naturelle. Livre XX. Texte établi, traduit et commenté par J. André.* Paris: Les Belles Lettres.

André, J. (1969). *Pline l'Ancien. Histoire naturelle. Livre XXI. Texte établi, traduit et commenté par J. André.* Paris: Les Belles Lettres.

André, J. (1970). *Pline l'Ancien. Histoire naturelle. Livre XXII. Texte établi, traduit et commenté par J. André.* Paris: Les Belles Lettres.

André, J. (1971). *Pline l'Ancien. Histoire naturelle. Livre XXIII. Texte établi, traduit et commenté par J. André.* Paris: Les Belles Lettres.

André, J. (1972). *Pline l'Ancient. Histoire naturelle. Livre XXIV. Texte établi, traduit et commenté par J. André.* Paris: Les Belles Lettres.

André, J. (1974). *Pline l'Ancien. Histoire naturelle. Livre XXV. Texte établi, traduit et commenté par J. André.* Paris: Les Belles Lettres.

André, J. (1981). *Étymologies. Livre XVII: De l'agriculture.* Paris: Les Belles Lettres.

André, J. (1985). *Les noms de plantes dans la Rome antique.* Paris: Les Belles Lettres.

Arber, A. (1912). *Herbals: Their Origin and Evolution. A Chapter in the History of Botany 1470–1670.* Cambridge: Cambridge University Press. Note: a revised edition (1970) exists, but we were interested in the original edition.

Armstrong, A. H. (1966). *Plotinus. Volume I. Porphyry on the Life of Plotinus and the Order of his Books. Enneads I.1–9. With an English Translation by A.H. Armstrong.* Cambridge, MA and London: Harvard University Press.

Ash, H. B. (1941). *Lucius Junius Moderatus Columella. On Agriculture I–IV. With an English Translation by H. B. Ash.* Cambridge, MA and London: Loeb Classical Library, Harvard University Press.

Atran, S. (1987). 'Origin of the species and genus concepts: an anthropological perspective'. *Journal of the History of Biology*, 20(2), 195–279.

Aufmesser, M. (2002). *Pedanius Dioscurides aus Anazarba: fünf Bücher über die Heilkunde. Aus dem griechischen übersetzt von M. Aufmesser*. Hildesheim: Olms-Weidmann.

Avanzini, A. (Ed.). (1997). *Profumi d'Arabia. Atti del convegno*. Rome: 'L'Erma' di Bretschneider.

Bakels, C. and Jacomet, S. (2003). 'Access to luxury foods in Central Europe during the Roman period: the archaeobotanical evidence'. *World Archaeology*, 34(3), 542–557.

Balme, D. M. (1962a). '*Genos* and *eidos* in Aristotle's biology'. *Classical Quarterly*, 12, 81–98.

Balme, D. M. (1962b). 'Development of biology in Aristotle and Theophrastus: theory of spontaneous generation'. *Phronesis*, 7, 91–104.

Balme, D. M. and Gotthelf, A. (1991). *Aristotle. History of Animals. Books VII–X. Edited and Tranlated by D.M. Balme. Prepared for Publication by A. Gotthelf.* Cambridge, MA and London: Harvard University Press.

Barbaud, J. (1994). 'Les Dioscorides "alphabétiques" (à propos du Codex Vindobonensis Med. Gr. 1)'. *Revue d'histoire de la pharmacie*, 302, 321–330.

Baumann, H. (1993). *Greek Wild Flowers and Plant Lore in Ancient Greece. Translated and augmented by W.T. Stearn and E.R. Stearn.* Third edition. London: Herbert Press.

Baumann, H. (2000). *Pflanzenbilder auf griechischen Münzen*. Munich: Hirmer.

Baumann, H. (2007). *Flora mythologica. Griechische Planzenwelt in der Antike*. Kilchberg and Zurich: Akanthus.

Beagon, M. (1992). *Roman Nature: The Thought of Pliny the Elder*. Oxford: Clarendon Press.

Beagon, M. (2013). '*Labores pro bono publico*. The burdensome mission of Pliny's *Natural history*'. In J. König and G. Woolf (Eds), *Encyclopaedism from Antiquity to the Renaissance* (pp. 84–107). Cambridge: Cambridge University Press.

Beatty, W. K. (1974). 'Medical numismatic notes, XV: some medical aspects of Greek and Roman coins'. *Bulletin of the New York Academy of Medicine*, 50(1), 85–95.

Beck, H. (1895). *Geoponica sive Cassiani Bassi scholastici de re rustica eclogue*. Leipzig: Teubner.

Beck, L. Y. (2005). *Pedanius Dioscorides of Anazarbus De materia medica. Translated by L.Y. Beck.* Hildesheim: Olms-Weidmann.

Bertoldi, V. (1936). 'Nomina tusca in Dioscoride'. *Studi Etruschi*, 10, 295–320.

Betz, H. D. (1992). *The Greek Magical Papyri in Translation, including the Demotic Spells. Volume one: Texts. With an updated Bibliography.* Second edition. Chicago and London: University of Chicago Press.

Billerbeck, H. L. J. (1824). *Flora classica*. Leipzig: Hinrichssche Buchhandlung.

Biraud, M. (1993). 'Usages de l'asphodèle et étymologies d'*asphodelos*'. In *Actes du colloque international 'Les phytonymes grecs et latins' tenu à Nice les 14, 15, et 16 mai 1992 à la Faculté des Lettres, Arts et Sciences humaines, Université de Nice-Sophia Antipolis* (pp. 35–46). Nice: Centre de recherches comparatives sur les langues de la Méditerranée ancienne.

Biville, F. (1993). 'La réception des hellénismes dans le lexique phytonymique latin (d'après Pline, *Histoire Naturelle*)'. In *Actes du colloque international 'Les phytonymes grecs et latins' tenu à Nice les 14, 15, et 16 mai 1992 à la Faculté des Lettres, Arts et Sciences humaines, Université de Nice-Sophia Antipolis* (pp. 47–61). Nice: Centre de recherches comparatives sur les langues de la Méditerranée ancienne.

Blanc, A. (1993). 'Les noms des parties constitutives de l'arbre en grec ancien: évolution et étymologie'. In *Actes du colloque international 'Les phytonymes grecs et latins' tenu à Nice les 14, 15, et 16 mai 1992 à la Faculté des Lettres, Arts et Sciences humaines, Université de Nice-Sophia Antipolis* (pp. 63–72). Nice: Centre de recherches comparatives sur les langues de la Méditerranée ancienne.

Blanc, A. (1996). 'Les ὀρεοτύποι de Théophraste: Carriers, bûcherons ou muletiers?' *Revue de philologie, de littérature et d'histoire anciennes*, 70(2), 199–210.

Blunt, W. and Stearn, W. T. (1994). *The Art of Botanical Illustration*. Revised edition. London: Antique Collectors' Club in association with the Royal Botanic Gardens, Kew.

Bodson, L. (1991). 'Alexander the Great and the scientific exploration of the oriental part of his empire: an overview of the background, trends and results'. *Ancient Society*, 22, 127–138.

Bologa, V. L. (1930). 'I sinonimi "daci" delle piante descritte da Dioscoride possono servire alla ricostruzione della lingua daca?' *Archeion*, 12(2), 166–170.

Bonet, V. (1993). 'Les noms des plantes médicales et la description du milieu naturel'. In *Actes du colloque international 'Les phytonymes grecs et latins' tenu à Nice les 14, 15, et 16 mai 1992 à la Faculté des Lettres, Arts et Sciences humaines, Université de Nice-Sophia* (pp. 73–83). Nice: Centre de recherches comparatives sur les langues de la Méditerranée ancienne.

Boudon-Millot, V. (2003). 'Aux marges de la médecine rationnelle: médecins et charlatans à Rome au temps de Galien (IIe s. de notre ère)'. *Revue des Etudes Grecques*, 116, 109–131.

Boudon-Millot, V. (2012). *Galien de Pergame: un médecin grec à Rome*. Paris: Les Belles Lettres.

Bovey, M. (1999). 'La greffe de l'olivier sur le figuier. Columelle, Virgile et la greffe des arbres'. *Revue des Etudes Latines*, 77, 184–204.

Bowe, P. (2004). *Gardens of the Roman World*. London: Frances Lincoln Ltd.

Braund, D. and Wilkins, J. (Eds). (2000). *Athenaeus and his World: Reading Greek Culture in the Roman Empire*. Exeter: Exeter University Press.

Bremmer, J. N. (1999). 'Paradise: from Persia, via Greece, into the Septuagint'. In G. P. Luttikhuizen (Ed.), *Paradise Interpreted: Representations of Biblical Paradise in Judaism and Christianity* (pp. 1–20). Leiden: Brill.

Bretin-Chabrol, M. and Leduc, C. (2009). 'La botanique antique et la problématique du genre'. *Clio. Histoire, femmes et sociétés*, 68(1), 205–223.

Bretzl, H. (1903). *Botanische Forschungen des Alexanderzuges*. Leipzig: Teubner.

Brosig, M. (1883). *Die Botanik des älteren Plinius*. Graudenz: Druk von Gustav Rōthe.

Brownson, C. L. (1918). *Xenophon. Hellenica. Books I–V. With an English Translation by C.L. Brownson*. London and Cambridge, MA: Loeb Classical Library, Harvard University Press.

Brownson, C. L. and Dillery, J. (1998). *Xenophon. Anabasis. Translated by C.L. Brownson and Revised by J. Dillery*. Cambridge, MA and London: Loeb Classical Library, Harvard University Press.

Brubaker, L. (2002). 'The Vienna Dioskorides and Anicia Juliana'. In A. Littlewood, H. Maguire and J. Wolschke-Bulmahn (Eds), *Byzantine Garden Culture* (pp. 189–214). Washington, DC: Dumbarton Oaks.

Brundrett, N. (2010). 'Roman tomb gardens: the construction of sacred commemorative landscapes'. *The Brock Review*, 11(2), 51–69.

Brunt, P. A. (1983). *Arrian. Anabasis of Alexander. Volume II. Books 5–7. Indica. With an English Translation by P.A. Brunt*. Cambridge, MA and London: Loeb Classical Library, Harvard University Press.

Buckheit, V. (1960). 'Feigensymbolik'. *Rheinisches Museum*, 103, 200–229.

Buller, A. R. (1914–1916). 'The fungus lore of the Greeks and Romans'. *Transactions of the British Mycological Society*, 5, 21–66.

Bury, R. G. (1929). *Plato. Timaeus. Critias. Cleitophon. Menexenus. Epistles. With an English Translation by the Reverend R.G. Bury*. London and Cambridge, MA: Loeb Classical Library, Harvard University Press.

Byl, S. (1980). *Recherches sur les grands traités biologiques d'Aristote: sources écrites et préjugés*. Brussels: Palais des académies.

Cadden, J. (1995). *The Meanings of Sex Difference in the Middle Ages: Medicine, Science, and Culture*. Cambridge: Cambridge University Press.

Calame, C. (2006). 'Jardins d'amour et prairies de l'au-delà: Rencontres rituelles avec les dieux et poésie en Grèce antique'. *Poétique*, 145, 25–41.

Calderón Dorda, E., De Lazzer, A. and Pellizer, E. (2003). *Fiumi e monti. Corpus Plutarchi Moralium 38*. Naples: M. D'Auria.

Campbell, J. B. (2012). 'Crowns and wreaths'. In S. Hornblower, A. Spawforth and E. Eidinow (Eds), *Oxford Classical Dictionary*. Fourth edition. Oxford: Oxford University Press.

Caneva, G. (2011). *Il codice botanico di Augusto. Roma, Ara Pacis: parlare al popolo attraverso le immagini della natura. The Augustus Botanical Code. Rome, Ara Pacis: Speaking to the People through the Images of Nature*. Rome: Gangemi Editore spa.

Capelle, W. (1954a). 'Theophrast in Kyrene'. *Rheinisches Museum*, 97, 169–187.

Capelle, W. (1954b). 'Der Garten des Theophrast'. In *Festschrift für Friedrich Zucker zum 70. Geburtstage* (pp. 48–82). Berlin: Akademie-Verlag.

Carnoy, A. J. (1959). *Dictionnaire étymologique des noms grecs de plantes*. Louvain: Publications universitaires.

Carroll-Spillecke, M. (1989). *Kepos: der antike griechische Garten*. Munich: Deutscher Kunstverlag.

Carroll-Spillecke, M. (Ed.). (1992). *Der Garten von der Antike bis zum Mittelalter*. Mainz am Rhein: P. von Zabern.

Carroll-Spillecke, M. (2003). *Earthly Paradises: Ancient Gardens in History and Archaeology*. London: The British Museum Press.

Casevitz, M. (1993). 'Sur les "mots voyageurs" de l'Inde jusqu'en Grèce'. *Topoi Orient-Occident*, 3, 401–406.

Cauvet, M. D. (1875). 'Sur le silphion'. *Bulletin de la Société botanique de France*, 22, 10–17.

Ciarallo, A. (2001). *Gardens of Pompeii. Translated by Lori-Ann Touchette*. Los Angeles: J. Paul Getty Museum.

Ciarallo, A. (2004). *Flora pompeiana*. Rome: 'L'Erma' di Bretschneider.

Cima, M. and La Rocca, E. (1998). *Horti Romani: atti del convegno internazionale Roma, 4–6 maggio 1995*. Rome: 'L'Erma' di Bretschneider.

Clement, P. A. and Hoffleit, H. B. (1969). *Plutarch's Moralia. Volume VIII. 612b-697c. With an English Translation by P.A. Clement and H.B. Hoffleit*. London and Cambridge, MA: Loeb Classical Library, Harvard University Press.

Clément-Tarantino, S. (2006). 'La poétique romaine comme hybridation féconde. Les leçons de la greffe (Virgile, Géorgiques, 2, 9–82)'. *Interférences Ars Scribendi*, 4, 1–26.

Collins, M. (2000). *Medieval Herbals: The Illustrative Traditions*. London: British Library.

Colson, F. H. and Whitaker, G. H. (1930). *Philo. On the Unchangeableness of God. On Husbandry. Concerning Noah's Work as a Planter. On Drunkenness. On Sobriety. With an English Translation by F.H. Colson and G.H. Whitaker*. Cambridge, MA and London: Loeb Classical Library, Harvard University Press.

Coubray, S., Zech-Matterne, V. and Mazurier, A. (2010). 'The earliest remains of a *Citrus* fruit from a western Mediterranean archaeological context? A microtomographic-based re-assessment'. *Comptes Rendus Palevol*, 9(6), 277–282.

Craik, E. M. (2014). *The 'Hippocratic' Corpus: Content and Context*. London: Routledge.

Crawford, D. J. (1973). 'Garlic-growing and agricultural specialization in Graeco-Roman Egypt'. *Chronique d'Egypte*, 48, 350–363.

Cronier, M. (2010). 'Le Dioscoride alphabétique latin et les traductions latines du *De materia medica*'. In D. Langslow and B. Maire (Eds), *Body, Disease and Treatment in a Changing World: Latin Texts and Contexts in Ancient and Medieval Medicine. Proceedings of the Ninth International Conference 'Ancient Latin Medical Texts', University of Manchester, 5th-8th September 2007* (pp. 189–200). Lausanne: Éditions BHMS.

Cunliffe, B. (1971). *Fishbourne: A Roman Palace and its Garden*. London: Thames & Hudson.

Cunliffe, B. (1981). 'Roman gardens in Britain: a review of the evidence'. In E. B. MacDougall and W. F. Jashemski (Eds), *Ancient Roman Gardens* (pp. 95–108). Washington, DC: Dumbarton Oaks.

Cusset, C. and Thibodeau, P. (2008). 'C. Iulius Hyginus'. In P. T. Keyser and G. L. Irby-Massie (Eds), *The Encyclopedia of Ancient Natural Scientists: The Greek Tradition and its Many Heirs* (pp. 454–455). London: Routledge.

Dalby, A. (2000a). *Dangerous Tastes: The Story of Spices*. London: British Museum Press.

Dalby, A. (2000b). 'To feed a king: tyrants, kings and the search for quality in agriculture and food'. *Pallas*, 52, 133–144.

Dalby, A. (2003). *Food in the Ancient World from A to Z*. London and New York: Routledge.

Dalby, A. (2011). *Geoponika: Farm Work. A Modern Translation of the Roman and Byzantine Farming Handbook*. Totnes: Prospect.

Daly, L. W. (1967). *Contributions to a History of Alphabetization in Antiquity and the Middle Ages*. Brussels: Latomus.

Davis, P. H., Cullen, J. and Coode, M. J. (1965–1988). *Flora of Turkey and the East Aegean Islands*. 10 vols. Edinburgh: Edinburgh University Press.

Dawkins, R. M. (1936). 'The semantics of Greek names for plants'. *Journal of Hellenic Studies*, 56(1), 1–11.

De Romanis, F. (1996). *Cassia, cinnamomo, ossidiana. Uomini e mercanti tra oceano indiano e Mediterraneo*. Rome: L' 'Erma' di Bretschneider.

Debru, A. (1996). 'The gardener and the lady: therapeutics and society in the age of Galen'. In R. Pötzsch (Ed.), *The Pharmacy: Windows on History* (pp. 23–33). Basel: Editiones Roche.

Debru, A. (Ed.). (1997). *Galen on Pharmacology: Philosophy, History, and Medicine: Proceedings of the Vth International Galen Colloquium, Lille, 16–18 March 1995*. Leiden: Brill.

Delano Smith, C. (1996). 'Where was the "wilderness" in Roman times?' In G. Shipley and J. Salmon (Eds), *Human Landscapes in Classical Antiquity: Environment and Culture* (pp. 154–179). London: Routledge.

Delatte, A. (1939). *Anecdota atheniensia et alia. Tome II. Textes grecs relatifs à l'histoire des sciences*. Liége: Faculté de philosophie et lettres.

Delatte, A. (1961). *Herbarius: recherches sur le cérémonial usité chez les anciens pour la cueillette des simples et des plantes magiques*. Third edition. Brussels: Palais des Académies.

della Cella, P. and Viviani, D. (1819). *Viaggio da Tripoli di Barberia alle frontiere occidentali dell'Egitto fatto nel 1817*. Genova: A. Ponthenier.

Denham, A. and Whitelegg, M. (2014). 'Deciphering Dioscorides: mountains and mole-hills?' In S. Francia and A. Stobart (Eds), *Critical Approaches to the History of Western Herbal Medicine: From Classical Antiquity to the Early Modern Period* (pp. 191–209). London: Bloomsbury.

Deroux, C. O. (1976). 'Note sur l'ellébore et le faux ellébore'. *Latomus*, 35, 875–878.

Desautels, J. (1988). 'La classification des végétaux dans la *Recherche sur les plantes* de Théophraste d'Erésos'. *Phoenix*, 42(3), 219–243.

Desautels, J. and Girard, M.-C. (1983). 'Les végétaux dans le *Corpus Hippocraticum*'. In F. Lasserre and P. Mudry (Eds), *Formes de pensée dans la collection hippocratique. Actes du 4e Colloque International Hippocratique, Lausanne, 21–26 Septembre, 1981* (pp. 187–201). Geneva: Droz.

Detienne, M. (1972). *Les jardins d'Adonis*. Paris: Gallimard.

Detsicas, A. (1981). 'A group of pottery from Eccles, Kent'. In A. C. Anderson and A. S. Anderson (Eds), *Roman Pottery Research in Britain and North-West Europe. Papers presented to Graham Webster. Part II* (pp. 441–445). Oxford: B.A.R.

Diederich, S. (2007). *Römische Agrarhandbücher zwischen Fachwissenschaft, Literatur und Ideologie*. Berlin: Walter de Gruyter.

Dierbach, J. H. (1824). *Die Arzneimittel des Hippokrates oder Versuch einer systematischen Aufzählung der in allen hippokratischen Schriften vorkommenden Medikamente*. Heidelberg: Karl Groos.

Dierbach, J. H. (1833). *Flora mythologica, oder Pflanzenkunde in Bezug auf Mythologie und Symbolik der Griechen und Römer. Ein Beitrag zur ältesten Geschichte der Botanik, Agricultur und Medicin*. Frankfurt am Main: Gedruckt und verlägt bei Johann David Sauerländer.

Dillon, M. P. (1997). 'The ecology of the Greek sanctuary'. *Zeitschrift für Papyrologie und Epigraphik*, 118, 113–127.

Dognini, C. (2008). 'Androsthenēs of Thasos'. In P. T. Keyser and G. L. Irby-Massie (Eds), *The Encyclopedia of Ancient Natural Scientists: The Greek Tradition and its Many Heirs* (p. 82). London: Routledge.

Doody, A. (2001). 'Finding facts in Pliny's encyclopaedia: the Summarium of the Natural History'. *Ramus*, 30(1), 1–22.

Doody, A. (2011). 'The science and aesthetics of names in the *Natural History*'. In R. Gibson and R. Morello (Eds), *Pliny the Elder: Themes and Contexts* (p. 113). Leiden: Brill.

Dorie, M. (1967). 'Les plantes magiques de l'Odyssée. Lotos et moly'. *Revue d'histoire de la pharmacie*, 55(195), 573–584.

Drossaart Lulofs, H. J. (1957). 'Aristotle's ΠΕΡΙ ΦΥΤΩΝ'. *Journal of Hellenic Studies*, 77(1), 75–80.

Drossaart Lulofs, H. J. and Poortman, E. L. J. (1989). *Nicolaus Damascenus De plantis: Five Translations*. Amsterdam: North-Holland Publishing Company.

DuBois, P. (1988). *Sowing the Body: Psychoanalysis and Ancient Representations of Women*. Chicago: University of Chicago Press.

Ducourthial, G. (2003). *Flore magique et astrologique de l'antiquité*. Paris: Belin.

Duhoux, Y. (1993). 'Les premiers phytonymes grecs: les données mycéniennes'. In *Actes du colloque international 'Les phytonymes grecs et latins' tenu à Nice les 14, 15, et 16 mai 1992 à la Faculté des Lettres, Arts et Sciences humaines, Université de Nice-Sophia Antipolis* (pp. 97–122). Nice: Centre de recherches comparatives sur les langues de la Méditerranée ancienne.

Edmonds, J. M. (1912). *Greek Bucolic Poets. Theocritus. Bion. Moschus. With an English Translation by J.M. Edmonds*. Cambridge, MA and London: Loeb Classical Library, Harvard University Press.

Eichberger, C., Sigl, M. and Rühfel, H. (2007). 'Trees and shrubs on classical Greek vases'. *Bocconea*, 21, 117–130.

Eigler, U. and Wöhrle, G. (1993). *Theophrast De odoribus: Edition, Übersetzung, Kommentar*. Stuttgart: Teubner.

Einarson, B. and Link, G. K. K. (1976–1990). *Theophrastus. De causis plantarum*. 3 vols. Cambridge, MA and London: Harvard University Press.

Englert, W. G. (2008). 'Dēmokritos of Abdēra'. In P. T. Keyser and G. L. Irby-Massie (Eds), *The Encyclopedia of Ancient Natural Scientists: The Greek Tradition and its Many Heirs* (pp. 235–236). London: Routledge.

Ernout, A. (1948). 'Frutex-Frutico'. *Revue belge de philologie et d'histoire*, 26(1–2), 85–92.

Ernout, A. (1949). *Pline l'Ancien. Histoire naturelle. Livre XII. Texte établi et traduit par A. Ernout*. Paris: Les Belles Lettres.

Ernout, A. (1956). *Pline l'Ancien. Histoire naturelle. Livre XIII. Texte établi, traduit et commenté par A. Ernout*. Paris: Les Belles Lettres.

Ernout, A. (1959). *Pline l'Ancien. Histoire naturelle. Livre XXVII. Texte établi, traduit et commenté par A. Ernout*. Paris: Les Belles Lettres.

Ernout, A. and Pépin, R. (1957). *Pline l'ancien. Histoire naturelle. Livre XVXI. Texte établi, traduit et commenté par A. Ernout et R. Pépin*. Paris: Les Belles Lettres.

Euchholz, J. B. (1848). *Flora Homerica*. Chelmno: W. Th. Lohde.

Everett, N. (2012). *The Alphabet of Galen*. Toronto: University of Toronto Press.

Fabricatti, E. (1993). 'Silphium in ancient art'. *Libyan Studies*, 24, 27–33.

Fabricius, C. (1972). *Galens Exzerpte aus älteren Pharmakologen*. Berlin: Walter de Gruyter.

Fairclough, H. R. and Goold, G. P. (1999). *Virgil. Eclogues. Georgics. Aeneid I–IV. With an English Translation by H.R. Fairclough. Revised by G.P. Goold*. Cambridge, MA and London: Loeb Classical Library, Harvard University Press.

Farrar, L. (1998). *Ancient Roman Gardens*. Stroud: Sutton Publishing.

Fausti, D. (1997). 'Ricerche sul lessico botanico dei papiri medici'. In A. Andorline (Ed.), *'Specimina' per il Corpus dei papiri Greci di medicina. Atti dell'Incontro di studio, Firenze, 28–29 marzo 1996* (pp. 83–105). Florence: Istituto papirologico 'G. Vitelli'.

Fausti, D. (1998). 'MP3 2095 erbario illustrato'. In *Estratto provvisorio dal Corpus dei Papiri Greci di Medicina. Fascicolo presentato in occasione del XXXII Congresso Internazionale di Papirologia (Firenze, 23–25 agosto 1998)* (pp. 43–58). Florence: Istituto Papirologico 'G. Vitelli'.

Fausti, D. (2004). 'Erbari illustrati su papiro e tradizione iconografica botanica'. In I. Andorlini (Ed.), *Testi medici su papiro. Atti del Seminario di Studio (Firenze, 3–4 giugno 2002)* (pp. 131–150). Florence: Istituto Papirologico 'G. Vitelli'.

Fée, A. L. A. (1822). *Flore de Virgile, ou, Nomenclature méthodique et critique des plantes, fruits et produits végétaux, mentionnés dans les ouvrages du prince des poëtes latins*. Paris: Didot.

Fée, A. L. A. (1832). *Flore de Théocrite et des autres bucoliques grecs*. Paris: Didot.

Fée, A. L. A. (1833). *Commentaires sur la botanique et la matière médicale de Pline*. Paris: C.-L.-F. Panckoucke.

Fellner, S. (1897). *Die homerische Flora*. Vienna: A. Hölder.

Ferraces Rodríguez, A. (1999). *Estudios sobre textos latinos de fitoterapia entre la antigüedad tardía y la alta edad media*. Coruña: Universidade da Coruña.

Fisher, N. R. E. (1976). *The Concept of Hybris in Greece from Homer to the Fourth century B.C.* DPhil, University of Oxford.

Fitch, J. G. (2013). *Palladius. Opus agriculturae: The Work of Farming*. Totnes: Prospect.

Fitzpatrick, A. P. (1991). 'Ex Radice Britanica'. *Britannia*, 22, 143–146.

Flemming, R. (2007). 'Galen's imperial order of knowledge'. In J. König and T. Whitmarsh (Eds), *Ordering Knowledge in the Roman Empire* (pp. 241–277). Cambridge: Cambridge University Press.

Flory, M. B. (1995). 'The symbolism of laurel in cameo portraits of Livia'. *Memoirs of the American Academy in Rome*, 40, 43–68.

Forster, E. S. (1936). 'Trees and plants in Homer'. *Classical Review*, 50(3), 97–104.

Forster, E. S. (1942). 'Trees and plants in Herodotus'. *Classical Review*, 56(2), 57–63.

Forster, E. S. and Heffner, E. H. (1954). *Columella on Agriculture. Books 5–9*. Cambridge, MA and London: Loeb Classical Library, Harvard University Press.

Forster, E. S. and Heffner, E. H. (1955). *Lucius Junius Moderatus Columella On Agriculture X–XII. On Trees. With an English Translation by E.S. Forster and E.H. Heffner*. Cambridge, MA and London: Loeb Classical Library, Harvard University Press.

Fortes Fortes, J. (1997). 'Hispanische Pflanzennamen im Pseudo-Dioskurides und Pseudo-Apuleius'. *Glotta*, 74, 1–11.

Foster, E. D. (1974). *The Manufacture and Trade of Mycenaean Perfumed Oil*. Yale University, Ann Arbor.

Foucaud, A. (1962). 'Lotus et lotophages: un petit problème d'histoire grecque'. *Revue d'histoire de la pharmacie*, 50(175), 422–424.

Foxhall, L. (1998). 'Natural sex: the attribution of sex and gender to plants in ancient Greece'. In L. Foxhall and J. Salmon (Eds), *Thinking Men: Masculinity and its Self-Representation in the Classical Tradition* (pp. 57–70). London and New York: Routledge.

Foxhall, L. (2007). *Olive Cultivation in Ancient Greece: Seeking the Ancient Economy*. Oxford: Oxford University Press.

Fraas, K. N. (1845). *Synopsis plantarum florae classicae: oder, Uebersichtliche Darstellung der in den klassischen Schriften der Griechen und Römer vorkommenden Pflanzen, nach autoptischer Untersuchung im Florengebiete, entworfen und nach Synonymen geordnet*. Munich: Verlag von E.A. Fleischmann.

Francissen, F. P. (1987). 'A century of scientific research on plants in Roman mural paintings (1879–1979)'. *Rivista di studi pompeiani*, 1(1), 111–122.

Frass, M. (2006). *Antike römische Gärten: soziale und wirtschaftliche Funktionen der Horti Romani*. Horn: F. Berger und Söhne.

French, R. (1994). *Ancient Natural History*. London: Routledge.

Friedrich, H.-V. (1968). *Thessalos von Tralles griechisch und lateinish*. Meisenheim am Glan: Verlag Anton Hain.

Gabriel, M. M. (1955). *Livia's Garden Room at Prima Porta*. New York: New York University Press.

Gaillard-Seux, P. (1999). 'Le "sang de basilic" chez Pline l'Ancien (NH, XXIX, 66): résine de genévrier ou hématite?' *L'Antiquité classique*, 68, 227–238.

Garnsey, P. D. A. (1999). *Food and Society in Classical Antiquity*. Cambridge: Cambridge University Press.

Gaselee, S. (1969). *Achilles Tatius. Leucippe and Clitophon. With an English Translation by S. Gaselee*. Cambridge, MA and London: Loeb Classical Library, Harvard University Press.

Gemmill, C. L. (1973a). 'The missing passage in Hort's translation of Theophrastus'. *Bulletin of the New York Academy of Medicine*, 49(2), 127–129.

Gemmill, C. L. (1973b). 'Medical numismatic notes. 8. Coins of Cyrene'. *Bulletin of the New York Academy of Medicine*, 49(1), 81–84.

Georgi, L. (1982). 'Poillination ecology of the date palm and fig tree: Herodotus 1.193.4–5'. *Classical Philology*, 77(3), 224–228.

Gest, H. (2002). 'History of the word photosynthesis and evolution of its definition'. *Photosynthesis Research*, 73(1–3), 7–10.

Giesecke, A. (2014). *The Mythology of Plants: Botanical Lore from Ancient Greece and Rome*. Los Angeles: Getty Publications.

Giet, S. (1950). *Basile de Césarée. Homélies sur l'Hexaéméron*. Paris: Les Editions du Cerf.

Girard, M.-C. (1988). *Connaissance et méconnaissance de l'hellébore dans l'Antiquité*. Québec: Laboratoire de recherches hippocratiques. Université Laval.

Girard, M.-C. (1990). 'L'hellébore: panacée ou placébo?' In P. Potter, G. Maloney and J. Desautels (Eds), *La maladie et les malades dans la Collection hippocratique: actes du VIe Colloque international hippocratique (Québec, du 28 septembre au 3 octobre 1987)* (pp. 393–405). Québec: Editions du Sphinx.

Gleason, K. L. (2010). 'Constructing nature: the built garden. With notice of a new monumental garden at the Villa Arianna, Stabiae'. *Bolletino di archeologia on line, Volume speciale*, 8–15.

Gotthelf, A. (1988). 'Historia I: plantarum et animalium'. In W. W. Fortenbaugh and R. W. Sharples (Eds), *Theophrastean Studies on Natural Science, Physics and Metaphysics, Ethics, Religion, and Rhetoric* (pp. 100–135). New Brunswick: Transaction Books.

Gotthelf, A. (1989). 'Teleology and spontaneous generation in Aristotle: a discussion'. *Apeiron*, 22(4), 181–194.

Gottschalk, H. B. (1998). 'Theophrastus and the Peripatos'. In J. M. van Ophuijsen and M. van Raalte (Eds), *Theophrastus, Reappraising the Sources* (pp. 281–298). New Brunswick and London: Transaction Publishers.

Gow, A. S. F. and Schofield, A. F. (1953). *Nicander of Colophon: Poems and Poetical Fragments*. Cambridge: Cambridge University Press.

Graham, D. W. (2008a). 'Anaxagoras of Klazomenai'. In P. T. Keyser and G. L. Irby-Massie (Eds), *The Encyclopedia of Ancient Natural Scientists: The Greek Tradition and its Many Heirs* (pp. 73–74). London: Routledge.

Graham, D. W. (2008b). 'Diogenēs of Apollōnia'. In P. T. Keyser and G. L. Irby-Massie (Eds), *The Encyclopedia of Ancient Natural Scientists: The Greek Tradition and its Many Heirs* (p. 252). London: Routledge.

Granger, H. (1985). 'The *scala naturae* and the continuity of kinds'. *Phronesis*, 30(2), 181–200.

Grant, M. (1997). *Dieting for an Emperor: A Translation of Books 1 and 4 of Oribasius' Medical Compilation with an Introduction and Commentary*. Leiden: Brill.

Grant, M. (2000). *Galen on Food and Diet*. London: Routledge.

Grape-Albers, H. (1977). *Spätantike Bilder aus der Welt des Arztes: medizinische Bilderhandschriften der Spätantike und ihre mittelalterliche Überlieferung*. Wiesbaden: Pressler.

Greene, E. L. (1909). *Landmarks of Botanical History: A Study of Certain Epochs in the Development of the Science of Botany. Part 1 – prior to 1562 AD*. Volume 1. Washington: Smithsonian Institute. Note: a revised edition (1983, edited by N. Egerton) exists, but we were interested in the original edition.

Grenfell, B. P. and Hunt, A. S. (1922). *The Oxyrhynchus Papyri. Part XV. Edited with Translations and Notes by B.P Grenfell and A.S. Hunt*. London and Oxford: Henry Frowde and Oxford University Press.

Grenfell, B. P., Hunt, A. S. and Goodspeed, E. J. (1907). *The Tebtunis Papyri*. Volume II. London: Egypt Exploration Society.

Grimal, P. (1984). *Les jardins romains. Essai sur le naturalisme romain*. Third edition. Paris: Fayard.

Grimm-Samuel, V. (1991). 'On the mushroom that deified the Emperor Claudius'. *Classical Quarterly*, 41(1), 178–182.

Groom, N. (1981). *Frankincense and Myrrh: A Study of the Arabian Incense Trade*. London and New York: Longman.

Grove, A. T. and Rackham, O. (2003). *The Nature of Mediterranean Europe: An Ecological History*. Second edition. New Haven and London: Yale University Press.

Guasparri, A. (2013). 'Explicit nomenclature and classification in Pliny's *Natural History* XXXII'. *Studies in History and Philosophy of Science Part A*, 44(3), 347–353.

Halácsy, E. v. (1901). *Conspectus Florae Graecae. 3 vols. and supplement*. Leipzig: Engelmann.

Hallum, B. (2008). 'Bōlos of Mendēs'. In P. T. Keyser and G. L. Irby-Massie (Eds), *The Encyclopedia of Ancient Natural Scientists: The Greek Tradition and its Many Heirs* (pp. 196–197). London: Routledge.

Hanson, A. E. (2001). 'Text and context for the illustrated herbal from Tebtunis'. In I. Andorlini (Ed.), *Atti del XXII Congresso Internazionale di Papirologia (Firenze, 23–29 agosto 1998)* (pp. 585–604). Florence: Istituto papirologico G. Vitelli.

Hanson, A. E. (2005). 'Greek medical papyri from the Fayum village of Tebtunis: patient involvement in a local health-care system?' In P. J. van der Eijk (Ed.), *Hippocrates in Context: Papers Read at the XIth International Hippocrates Colloquium, University of Newcastle upon Tyne, 27–31 August 2002*. Volume 31 (pp. 387–402). Leiden: Brill.

Harris, W. V. (1991). *Ancient Literacy*. Cambridge, MA and London: Harvard University Press.

Harvey-Gibson, R. J. (1919). *Outlines of the History of Botany*. London: A. and C. Black.

Heiberg, J. L. (1924). *Paulus Aegineta. Pars altera, Libri V–VII. Edidit I.L. Heiberg. CMG IV 2*. Leipzig and Berlin: Teubner.

Heitsch, E. (1964). *Die griechischen Dichterfragmente der römischen Kaiserzeit. Band 2*. Göttingen: Vandenhoeck and Ruprecht.

Henderson, Jeffrey (1991). *The Maculate Muse: Obscene Language in Attic Comedy*. New York and Oxford: Oxford University Press.

Henderson, Jeffrey (1998). *Aristophanes. Volume I. Acharnians. Knights. Edited and Translated by J. Henderson.* Cambridge, MA and London: Loeb Classical Library, Harvard University Press.

Henderson, John (2002). 'Columella's living hedge: the Roman gardening book'. *Journal of Roman Studies*, 92, 110–133.

Henderson, John (2004). *The Roman Book of Gardening*. London: Routledge.

Hendrickson, G. L. and Hubbekk, H. M. (1962). *Cicero. Brutus. With an English Translation by G.L. Hendrickson. Orator. With an English Translation by H.M. Hubbell.* Revised edition. London and Cambridge, MA: Loeb Classical Library, Harvard University Press.

Herzhoff, B. (1984). 'Lotos: Botanische Beobachtungen zu einem homerischen Pflanzennamen'. *Hermes*, 112(3), 257–271.

Herzhoff, B. (2006). 'Ist die Schrift "*De Plantis*" von Aristoteles?' In J. Althoff, B. Herzhoff and G. Wörhle (Eds), *Antike Naturwissenschaft und ihre Rezeption. Band 16'* (pp. 69–108). Trier: Wissenschaftlicher Verlag.

Hett, W. S. (1931). *Aristotle. Minor Works. On Colours. On Things Heard. Physiognomics. On Plants. On Marvellous Things Heard. Mechanical Problems. On Indivisible Lines. Situations and Names of Winds. On Melissus, Xenophanes and Gorgias. With an English Translation by W.S. Hett.* London and Cambridge, MA: Loeb Classical Library, William Heinemann and Harvard University Press.

Hett, W. S. (1957). *Aristotle. On the Soul. Parva Naturalia. On Breath.* Cambridge, MA and London: Harvard University Press.

Hicks, R. D. (1925). *Diogenes Laertius. Lives of Eminent Philosophers. Volume I. With an English Translation by R.D. Hicks*. London and Cambridge, MA: Loeb Classical Library, Harvard University Press.

Hooper, W. D. and Ash, H. B. (1934). *Marcus Porcius Cato. On Agriculture. Marcus Terentius Varro. On Agriculture*. Cambridge, MA and London: Loeb Classical Library, Harvard University Press.

Horden, P. and Purcell, N. (2000). *The Corrupting Sea: A Study of Mediterranean History*. Oxford: Blackwell.

Hort, A. F. (1916). *Theophrastus. Enquiry into Plants. Volume I. Books 1–5.* Cambridge, MA and London: Loeb Classical Library, Harvard University Press.

Hort, A. F. (1926). *Theophrastus. Enquiry into Plants. Volume II. Books 6–9. Minor Works on Odours and Weather Signs.* Cambridge, MA and London: Loeb Classical Library, Harvard University Press.

Houghton, W. (1885). 'Notices of fungi in Greek and Latin authors'. *Annals and Magazine of Natural History*, 15(85), 22–49.

Howald, E. and Sigerist, H. E. (1927). *Antonii Musae de herba vettonica liber. Pseudo Apulei Herbarius. Anonymi de taxone liber. Sexti Placiti liber medicinae ex animalibus*. Leipzig and Berlin: Teubner.

Hughes, J. D. (1983). 'How the ancients viewed deforestation'. *Journal of Field Archaeology*, 10(4), 437–445.

Hughes, J. D. (1988). 'Theophrastus as ecologist'. In W. W. Fortenbaugh and R. W. Sharples (Eds), *Theophrastean Studies on Natural Science, Physics and Metaphysics, Ethics, Religion, and Rhetoric* (pp. 67–75). New Brunswick: Trancation Books.

Hughes, J. D. (2006). 'The natural environment'. In K. H. Kinzl (Ed.), *A Companion to the Classical Greek World* (pp. 227–244). London: Wiley-Blackwell.

Hughes, J. D. (2011). 'Ancient deforestation revisited'. *Journal of the History of Biology*, 44, 43–57.

Hughes, J. D. and Thirgood, J. V. (1982). 'Deforestation, erosion, and forest management in ancient Greece and Rome'. *Journal of Forest History*, 26(2), 60–75.

Hünermörder, C. (2010). 'Zoology and botany'. In H. Cancik and S. Helmut (Eds), *Brill's New Pauly Encyclopedia of the Ancient World. Antiquity. Vol. 15. TUC-ZYT. Addenda* (pp. 955–962). Leiden and Boston.

Imhoof-Blumer, F. and Keller, O. (1889). *Tier-und Pflanzenbilder auf Münzen und Gemmen des klassischen Alterums*. Leipzig: Teubner.

Irby-Massie, G. L. (2008a). 'Andreas of Karustos'. In P. T. Keyser and G. L. Irby-Massie (Eds), *The Encyclopedia of Ancient Natural Scientists: The Greek Tradition and its Many Heirs* (pp. 77–78). London: Routledge.

Irby-Massie, G. L. (2008b). 'Kleëmporos'. In P. T. Keyser and G. L. Irby-Massie (Eds), *The Encyclopedia of Ancient Natural Scientists: The Greek Tradition and its Many Heirs* (p. 477). London: Routledge.

Irby-Massie, G. L. (2008c). 'Mētrodōros (Pharm.)'. In P. T. Keyser and G. L. Irby-Massie (Eds), *The Encyclopedia of Ancient Natural Scientists: The Greek Tradition and its Many Heirs* (p. 553). London: Routledge.

Jacomet, S., Kučan, D., Ritter, A., Suter, G. and Hagendorn, A. (2002). '*Punica granatum* L. (pomegranates) from early Roman contexts in Vindonissa (Switzerland)'. *Vegetation History and Archaeobotany*, 11(1–2), 79–92.

Jacques, J.-M. (1979). 'Nicandre de Colophon poète et médecin'. *Ktèma*, 4, 133–149.

Jacques, J.-M. (2002). *Nicandre de Colophon. Œuvres. Tome II. Les thériaques, Fragments iologiques antérieurs à Nicandre. Texte établi et traduit par J.-M. Jacques.* Paris: Les Belles Lettres.

Jacques, J.-M. (2007). *Oeuvres. Tome 3. Les Alexipharmaques, lieux parallèles du livre XIII des Iatrica d'Aétius. Texte établi et traduit par J.-M. Jacques*. Paris: Les Belles Lettres.

Jacques, J.-M. (2008a). 'Apollodōros the thēriakos'. In P. T. Keyser and G. L. Irby-Massie (Eds), *The Encyclopedia of Ancient Natural Scientists: The Greek Tradition and its Many Heirs* (p. 106). London: Routledge.

Jacques, J.-M. (2008b). 'Iollas of Bithunia'. In P. T. Keyser and G. L. Irby-Massie (Eds), *The Encyclopedia of Ancient Natural Scientists: The Greek Tradition and its Many Heirs* (pp. 439–440). London: Routledge.

Jacques, J.-M. (2008c). 'Krateuas'. In P. T. Keyser and G. L. Irby-Massie (Eds), *The Encyclopedia of Ancient Natural Scientists: The Greek Tradition and its Many Heirs* (p. 491). London: Routledge.

Jacques, J.-M. (2008d). 'Nikandros (Nicander) of Kolophōn'. In P. T. Keyser and G. L. Irby-Massie (Eds), *The Encyclopedia of Ancient Natural Scientists: The Greek Tradition and its Many Heirs* (pp. 573–574). London: Routledge.

Jashemski, W. F. (1970). 'Tomb gardens at Pompeii'. *Classical Journal*, 66(2), 97–115.

Jashemski, W. F. (1974). 'The discovery of a market-garden orchard at Pompeii: the garden of the House of the Ship Europa'. *American Journal of Archaeology*, 78(4), 391–404.

Jashemski, W. F. (1979). *The Gardens of Pompeii, Herculaneum and the Villas Destroyed by Vesuvius*. New Rochelle: Caratzas Brothers.

Jashemski, W. F. (1981). 'Introduction'. In E. B. MacDougall and W. F. Jashemski (Eds), *Ancient Roman Gardens* (pp. 1–6). Washington, DC: Dumbarton Oaks.

Jashemski, W. F. (1992). 'Vasa Fictilia: Ollae Perforatae'. In R. M. Wilhelm and H. Jones (Eds), *Two Worlds of the Poet: New Perspectives on Vergil* (pp. 371–391). Detroit: Wayne State University Press.

Jashemski, W. F. (1996). 'Ancient Roman gardens in Campania and Tunisia: a comparison of the evidence'. *The Journal of Garden History*, 16(4), 231–243.

Jashemski, W. F. (1999). *A Pompeian Herbal*. Austin: University of Texas Press.

Jashemski, W. F. and Meyer, F. G. (Eds). (2002). *The Natural History of Pompeii*. Cambridge: Cambridge University Press.

Jashemski, W. F., Meyer, F. G. and Ricciardi, M. (2002). 'Plants: evidence from wall paintings, mosaics, sculpture, plants remains, graffiti, inscriptions, and ancient authors'. In W. F. Jashemski and F. G. Meyer (Eds), *The Natural History of Pompeii* (pp. 80–180). Cambridge: Cambridge University Press.

Jashemski, W. F., Ricotti, E. S. P. and Foss, J. (1992). 'Preliminary excavations in the gardens of Hadrian's Villa: the Canopus area and the Piazza d'Oro'. *American Journal of Archaeology*, 96(4), 579–597.

Jashemski, W. F., Foss, J. E., Lewis, R. J., Timpson, M. E. and Lee, S. (1995). 'Roman gardens in Tunisia: preliminary excavations in the House of Bacchus and Ariadne and in the East Temple at Thuburbo Maius'. *American Journal of Archaeology*, 99(4), 559–576.

Johnson, J. de M. (1913). 'A botanical papyrus with illustrations'. *Archive für Geschichte der Naturwissenschaften und der Technik*, 4, 403–408.

Johnson, K. P. (2006). 'An Etruscan herbal?' *Etruscan News*, 5, 1 and 8.

Jones, C. P. (2005). *Life of Apollonius of Tyana. Volume I. Books 1–4. Edited and Translated by C.P. Jones*. Cambridge, MA and London: Loeb Classical Library, Harvard University Press.

Jones, P. M. (1999). *Medicina antiqua: Codex Vindobonensis 93, Vienna, Österreichische Nationalbibliothek*. London: Harvey Miller.

Jones, W. H. S. (1918). *Pausanias. Description of Greece. Volume I. Books 1–2 (Attica and Corinth). With an English Translation by W.H.S. Jones*. Cambridge, MA and London: Loeb Classical Library, Harvard University Press.

Jones, W. H. S. (1923). *Hippocrates. Vol. I. With an English Translation by W.H.S. Jones*. London and Cambridge, MA: Loeb Classical Library, Harvard University Press.

Jones, W. H. S. (1931). *Hippocrates. Vol. IV. Heracleitus on the Universe. With an English Translation by W.H.S. Jones*. Cambridge, MA and London: Loeb Classical Library, Harvard University Press.

Jones, W. H. S. (1969). *Pliny. Natural History. Books 20–23. With an English Translation by W.H.S. Jones*. Revised edition. Cambridge, MA and London: Loeb Classical Library, Harvard University Press.

Jones, W. H. S. (1980). *Pliny. Natural History. Books XXIV–XXVII. With an English Translation by W.H.S. Jones*. Second edition. Cambridge, MA and London: Loeb Classical Library, Harvard University Press.

Joret, C. (1904). 'Les recherches botaniques de l'expédition d'Alexandre'. *Journal des Savants*, 498–504 and 611–620.

Jouanna, J. (1999). *Hippocrates. Translated from the French by M.B. DeBevoise*. Baltimore and London: Johns Hopkins.

Kaimakis, D. (1976). *Die Kyraniden*. Meisenheim am Glan: Verlag Anton Hain.

Kanngiesser, F. (1910). *Die flora des Herodot*. Leipzig: Vogel.

Kaplan, P. (2008a). 'Aristoboulos of Kassandreia'. In P. T. Keyser and G. L. Irby-Massie (Eds), *The Encyclopedia of Ancient Natural Scientists: The Greek Tradition and its Many Heirs* (p. 136). London: Routledge.

Kaplan, P. (2008b). 'Nearkhos of Crete'. In P. T. Keyser and G. L. Irby-Massie (Eds),

The Encyclopedia of Ancient Natural Scientists: The Greek Tradition and its Many Heirs (pp. 568–569). London: Routledge.

Kaplan, P. (2008c). 'Onēsikritos of Astupalaia'. In P. T. Keyser and G. L. Irby-Massie (Eds), *The Encyclopedia of Ancient Natural Scientists: The Greek Tradition and its Many Heirs* (pp. 591–592). London: Routledge.

Karamanolis, G. (2008a). 'Plōtinos'. In P. T. Keyser and G. L. Irby-Massie (Eds), *The Encyclopedia of Ancient Natural Scientists: The Greek Tradition and its Many Heirs* (pp. 672–674). London: Routledge.

Karamanolis, G. (2008b). 'Porphurios of Tyre'. In P. T. Keyser and G. L. Irby-Massie (Eds), *The Encyclopedia of Ancient Natural Scientists: The Greek Tradition and its Many Heirs* (pp. 688–689). London: Routledge.

Karttunen, K. (1997). *India and the Hellenistic World*. Helsinki: Finnish Oriental Society.

Kassel, R. and Austin, C. (1986). *Poetae comici graeci. Ediderunt R. Kassel et C. Austin. Vol. V. Damoxenus-Magnes*. Berlin and New York: De Gruyter.

Kassel, R. and Austin, C. (1991). *Poetae comici graeci. Ediderunt R. Kassel et C. Austin. Vol. III. Agathenor-Aristogenus*. Berlin and New York: De Gruyter.

Kästner, H. (1896). 'Pseudo-Dioscoridis de herbis femininis'. *Hermes*, 31(3), 578–636.

Kellum, B. A. (1994). 'The construction of landscape in Augustan Rome: the garden room at the Villa ad Gallinas'. *The Art Bulletin*, 76(2), 211–224.

Kenawi, M., Macaulay-Lewis, E. and McKenzie, J. (2012). 'A commercial nursery near Abu Hummus (Egypt) and re-use of amphoras for the trade in plants'. *Journal of Roman Archaeology*, 25(1), 195–226.

Keyser, P. T. (2008a). 'Antonius Castor'. In P. T. Keyser and G. L. Irby-Massie (Eds), *The Encyclopedia of Ancient Natural Scientists: The Greek Tradition and its Many Heirs* (p. 100). London: Routledge.

Keyser, P. T. (2008b). 'Nestōr of Laranda'. In P. T. Keyser and G. L. Irby-Massie (Eds), *The Encyclopedia of Ancient Natural Scientists: The Greek Tradition and its Many Heirs* (p. 572). London: Routledge.

Keyser, P. T. (2008c). 'Pamphilos of Alexandria'. In P. T. Keyser and G. L. Irby-Massie (Eds), *The Encyclopedia of Ancient Natural Scientists: The Greek Tradition and its Many Heirs* (pp. 606–607). London: Routledge.

Keyser, P. T. (2008d). 'Phainias of Eresos'. In P. T. Keyser and G. L. Irby-Massie (Eds), *The Encyclopedia of Ancient Natural Scientists: The Greek Tradition and its Many Heirs* (p. 641). London: Routledge.

Kiilerich, B. (2001). 'The image of Anicia Juliana in the Vienna Dioscurides: flattery or apropriation of imperial imagery?' *Symbolae Osloenses*, *76*(1), 169–190.

King, H. (1994). 'Sowing the field: Greek and Roman sexology'. In R. Porter and M. Teich (Eds), *Sexual Knowledge, Sexual Science: The History of Attitudes to Sexuality* (pp. 29–46). Cambridge: Cambridge University Press.

King, H. (1998). *Hippocrates' Woman. Reading the Female Body in Ancient Greece*. London: Routledge.

Kircher-Durand, C. (1993). 'La dérivation suffixale dans les désignations de vignes à partir du livre XIV de l'*Histoire naturelle* de Pline'. In *Actes du colloque international 'Les phytonymes grecs et latins' tenu à Nice les 14, 15, et 16 mai 1992 à la Faculté des Lettres, Arts et Sciences humaines, Université de Nice-Sophia Antipolis* (pp. 207–222). Nice: Centre de recherches comparatives sur les langues de la Méditerranée ancienne.

Klynne, A. and Liljenstolpe, P. (2000). 'Investigating the gardens of the Villa of Livia'. *Journal of Roman Archaeology*, 13(1), 220–233.

Koerper, H. and Kolls, A. (1999). 'The silphium motif adorning ancient Libyan coinage: marketing a medicinal plant'. *Economic Botany*, 53(2), 133–143.

Korpela, J. (1995). '*Aromatarii, pharmacopolae, thurarii et ceteri.* Zur Sozialgeschichte Roms'. In P. J. van der Eijk, H. F. J. Horstmanshoff and P. H. Schrijvers (Eds), *Ancient Medicine in its Socio-Cultural Context: Papers Read at the Congress Held at Leiden Universiy, 13–15 April 1992* (pp. 101–118). Amsterdam: Rodopi.

Kraak, W. K. (1942). 'Aristote est-il toujours resté fidèle à sa conception que la plante se tient la tête en bas?' *Mnemosyne*, 10(4), 251–262.

Krapinger, G. (2005). *[Quintilian]. Die Bienen des armen Mannes. Grössere Deklamationen 13.* Cassino: Edizioni dell'Università degli Studi di Cassino.

Kronenberg, L. (2009). *Allegories of Farming from Greece and Rome: Philosophical Satire in Xenophon, Varro and Virgil.* Cambridge: Cambridge University Press.

Kühn, K. G. (1821–1833). *Claudii Galeni opera Omnia.* 20 vols. Berlin and Leipzig: Libraria Car. Cnoblochii.

Laks, A. and Most, G. W. (1993). *Théophraste. Métaphysique. Texte établi et traduit par A. Laks et G.W. Most.* Paris: Les Belles Lettres.

Laronde, A. (1996). 'Le silphium sur les monnaies de Cyrène'. *Studi Miscellanei*, 29, 157–168.

Lateiner, D. (1989). *The Historical Method of Herodotus.* Toronto: University of Toronto Press.

Le Bonniec, H. and Le Boeuffle, A. (1972). *Pline l'Ancien. Histoire Naturelle. Livre XVIII. Texte établi, traduit et commenté par H. Le Bonniec avec la collaboration de A. Le Boeuffle.* Paris: Les Belles Lettres.

Lecocq, F. (2009). 'L'œuf du phénix. Myrrhe, encens et cannelle dans le mythe du phénix'. *Hérodote*, 2, 73.

Lee, H. D. P. (1952). *Aristotle. Volume VII. Meteorologica. With an English Translation by H.D.P. Lee.* Cambridge, MA and London: Loeb Classical Library, Harvard University Press.

Leith, D. (2006). 'The Antinoopolis illustrated herbal (PJohnson + PAntin. 3.124 = MP3 2095)'. *Zeitschrift für Papyrologie und Epigraphik*, 156, 141–156.

Lennox, J. G. (1980). 'Aristotle on *genera*, species and the "more and the less"'. *Journal of the History of Biology*, 13(2), 321–346.

Lennox, J. G. (1981). 'Teleolgy, chance, and Aristotle's theory of spontaneous generation'. *Journal of the History of Philosophy*, 20, 219–238.

Lennox, J. G. (1985). 'Theophrastus on the limits of teleology'. In W. W. Fortenbaugh, P. Huby and A. A. Long (Eds), *Theophrastus of Eresus: On His Life and Works* (pp. 143–163). New Brunswick: Transaction Books.

Leone, P. L. M. (1968). *Ioannis Tzetzae Historiae.* Naples: Libreria Scientifica Editrice.

Liddell, H. G., Scott, R., Jones, H. S. and McKenzie, R. (1996). *A Greek-English Lexicon.* Ninth edition. Oxford: Clarendon Press.

Lindsell, A. (1937). 'Was Theocritus a botanist?' *Greece and Rome*, 6(17), 78–93.

Linnaeus, C. (1755). *Philosophia botanica: in qua explicantur fundamenta botanica cum definitionibus partium, exemplis terminorum, observationibus rariorum, adiectis figuris aeneis.* Vienna: Trattner.

Littlewood, A., Maguire, H. and Wolschke-Bulmahn, J. (2002). *Byzantine Garden Culture.* Washington, DC: Dumbarton Oaks.

Littré, É. (1839–1861). *Oeuvres complètes d'Hipocrate.* 10 vols. Paris: J. B. Bailliére.

Lloyd, A. C. (1962). 'Genus, species and ordered series in Aristotle'. *Phronesis*, 7, 67–90.

Lloyd, G. E. R. (1966). *Polarity and Analogy: Two Types of Argumentation in Early Greek Thought*. Cambridge: Cambridge University Press.

Lloyd, G. E. R. (1979). *Magic, Reason and Experience: Studies in the Origin and Development of Greek Science*. London: Bristol Classical Press.

Lloyd, G. E. R. (1983). *Science, Folklore, and Ideology: Studies in the Life Sciences in Ancient Greece*. Cambridge: Cambridge University Press.

Lloyd, G. E. R. (1996). *Aristotelian Explorations*. Cambridge: Cambridge University Press.

Lonie, I. M. (1969). 'On the botanical excursus in *De natura pueri* 22–27'. *Hermes*, 97(4), 391–411.

Lonie, I. M. (1981). *The Hippocratic Treatises 'On Generation', 'On the Nature of the Child'; 'Diseases IV'. A Commentary by I.M. Lonie*. Berlin and New York: Walter de Gruyter.

Louis, P. (1993). *Aristote. Problèmes. Tome II. Sections X à XXVII. Texte établi et traduit par P. Louis*. Paris: Les Belles Lettres.

Lowe, D. (2010). 'The symbolic value of grafting in ancient Rome'. *Transactions of the American Philological Association*, 140(2), 461–488.

Luccioni, P. (2006). 'Ce que dit le poète à propos des fleurs'. In C. Cusset (Ed.), *Musa docta. Recherches sur la poésie scientifique dans l'Antiquité* (pp. 259–274). Saint-Etienne: Publications de l'université de Saint-Etienne.

Mabberley, D. J. (2008). *Mabberley's Plant-Book: A Portable Dictionary of Plants, their Classification and Uses*. Third edition. Cambridge: Cambridge University Press.

Macaulay-Lewis, E. R. (2006a). 'The role of *ollae perforatae* in understanding horticulture, planting techniques, garden design, and plant trade in the Roman world'. In J. P. Morel, J. T. Juan and J. C. Matamala (Eds), *The Archaeology of Crop Fields and Gardens: Proceedings from 1st Conference on Crop Fields and Gardens Archaeology, University of Barcelona (Barcelona, Spain, June 1–3rd 2006)* (pp. 207–220). San Spirito: EdiPuglia.

Macaulay-Lewis, E. R. (2006b). 'Planting pots at Petra: a preliminary study of *ollae perforatae* at the Petra Garden pool complex and at the "Great Temple"'. *Levant*, 38, 159–170.

Macaulay-Lewis, E. R. (2008). 'The fruits of victory: generals, plants, and power in the Roman world'. In E. Bragg, L. I. Hau and E. R. Macaulay-Lewis (Eds), *Beyond the Battlefields: New Perspectives on Warfare and Society in the Graeco-Roman World* (pp. 205–225). Cambridge: Cambridge Scholars Publishing.

Macaulay-Lewis, E. R. (2010). 'Imported exotica: approaches to the study of the ancient plant trade'. *Bolletino di archeologia on line*, 1, 16–26.

McCabe, A. and Rodgers, R. H. (2008). 'Vindonius Anatolios of Bērutos'. In P. T. Keyser and G. L. Irby-Massie (Eds), *The Encyclopedia of Ancient Natural Scientists: The Greek Tradition and its Many Heirs* (pp. 72–73). London: Routledge.

MacDougall, E. B. and Jashemski, W. F. (1981). *Ancient Roman Gardens*. Washington, DC: Dumbarton Oaks.

McEnerney, J. I. (1983). 'Precatio Terrae and Precatio omnium herbarum'. *Rheinisches Museum für Philologie*, 126, 175–187.

McNamara, L. (2003). 'Conjurers, purifiers, vagabonds and quacks? The clinical roles of the folk and Hippocratic healers of classical Greece'. *Iris: Journal of the Classical Association of Victoria*, 16–17, 2–25.

Mägdefrau, K. (1973). *Geschichte der Botanik. Leben und Lesitung großer Forscher.* Stuttgart: Gustav Fischer Verlag.

Maggiulli, G. (1977). *Nomenclatura micologica latina.* Genova: Facoltà di lettere.

Maggiulli, G. and Buffa Giolito, M. F. (1996). *L'altro Apuleio: problemi aperti per una nuova edizione dell'Herbarius.* Naples: Loffredo.

Maire, B. (2002). *Les remèdes tirés des légumes et des fruits.* Paris: Les Belles Lettres.

Maloney, G. (1989). 'Les végétaux dans les livres des *Epidémies*'. In G. Baader and R. Winau (Eds), *Die Hippokratischen Epidemien, Théorie, Praxis, Tradition. Verhandlungen des Ve Colloque International Hippocratique, veranstaltelt von der Berliner Gesellschaft für Geschichte der Medizin in Verbindung mit dem Institut für Geschichte der Medizin der Freien Universität Berlin, 10–15.9.1984* (pp. 166–181). Stuttgart: F. Steiner.

Manetti, D. (2008). 'Dioklēs of Karustos'. In P. T. Keyser and G. L. Irby-Massie (Eds), *The Encyclopedia of Ancient Natural Scientists: The Greek Tradition and its Many Heirs* (pp. 254–255). London: Routledge.

Manunta, A. (2002). 'Cachrys ferulacca (L.) Calestani e il "silfio cirenaico"? Identificazione botanica su basi teoriche e rappresentazioni su monete'. *Quaderni di Archeologia della Libya*, 16, 345–350.

Marasco, G. (1996). 'Les médecins de cour à l'époque hellénistique'. *Revue des Etudes Grecques*, 109(2), 435–466.

Marganne, M.-H. (1992). 'Les références à l'Égypte dans la Matière médicale de Dioscoride'. In *Serta Leodiensia Secunda: Mélanges publiés par les Classiques de Liège à l'occasion du 175e anniversaire de l'Université* (pp. 309–322). Liège: C.I.P.L.

Marganne, M.-H. (1996). 'Un cinnamone fantomatique'. *Topoi*, 6, 657–664.

Marganne, M.-H. (2004). *Le livre médical dans le monde gréco-romain. Préface de Danielle Gourevitch.* Liège: CeDoPal.

Marshall, L. B. (1918). *L'horticulture antique et le poème de Columelle (De re rustica, livre X).* Paris: Hachette.

Marsilio, M. (2008a). 'Kleidēmos of Athens'. In P. T. Keyser and G. L. Irby-Massie (Eds), *The Encyclopedia of Ancient Natural Scientists: The Greek Tradition and its Many Heirs* (p. 478). London: Routledge.

Marsilio, M. (2008b). 'Leophanēs'. In P. T. Keyser and G. L. Irby-Massie (Eds), *The Encyclopedia of Ancient Natural Scientists: The Greek Tradition and its Many Heirs* (p. 504). London: Routledge.

Martyn, J. and King, E. (1755). *Publii Virgilii Maronis Georgicorum libri quatuor: The Georgicks of Virgil, with an English Translation and Notes.* London: Printed for T. Osborne and J. Shipton in Gray's-Inn.

Massar, N. (2005). *Soigner et servir: histoire sociale et culturelle de la médecine grecque à l'époque hellénistique.* Paris: De Boccard.

Masson, E. (1967). *Recherches sur les plus anciens emprunts sémitiques en grec.* Paris: C. Klincksieck.

Mayhew, R. (2010). *The Female in Aristotle's Biology: Reason or Rationalization.* Chicago: University of Chicago Press.

Mayhew, R. (2011). *Aristotle. Problems. Books 1–19. Edited and Translated by R. Mayhew.* Cambridge, MA and London: Loeb Classical Library, Harvard University Press.

Mayhew, R. and Mirhady, D. C. (2011). *Aristotle. Problems. Books 20–38. Edited and Translated by R. Mayhew. Rhetoric to Alexander. Edited and Translated by D.C.*

Mirhady. Cambridge, MA and London: Loeb Classical Library, Harvard University Press.

Mazzini, I. (1978 [second edition 1988]). *Q. Gargilii Martialis De hortis*. Bologna: Pàtron editore.

Megaloudi, F. (2005). 'Burnt sacrificial plant offerings in Hellenistic times: an archaeobotanical case study from Messene, Peloponnese, Greece'. *Vegetation History and Archaeobotany*, 14(4), 329–340.

Megaloudi, F., Papadopoulos, S. and Sgourou, M. (2007). 'Plant offerings from the classical necropolis of Limenas, Thasos, northern Greece'. *Antiquity*, 81, 933–943.

Meiggs, R. (1982). *Trees and Timber in the Ancient Mediterranean World*. Oxford: Clarendon Press.

Meliadò, C. (2008). 'Nikias of Miletus'. In P. T. Keyser and G. L. Irby-Massie (Eds), *The Encyclopedia of Ancient Natural Scientists: The Greek Tradition and its Many Heirs* (pp. 576–577). London: Routledge.

Messineo, G. (1984). 'Ollae perforatae'. *Xenia*, 8, 65–82.

Meyer, A. E. (1841). *Nicolai Damasceni De plantis libri duo Aristoteli vulgo adscripti: ex Isaaci Ben Honain versione Arabica Latine vertit Alfredus; Ad. codd. mss. fidem addito apparatu critico*. Leipzig: L. Voss.

Meyer, E. H. F. (1854–1857). *Geschichte der Botanik*. 4 vols. Königsberg: Verlag der Gebrüder Bornträger.

Michelini, A. (1978). ''Ύβρις and plants'. *Harvard Studies in Classical Philology*, 82, 35–44.

Miller, F. J. (1916). *Ovid. Metamorphoses. Volume I. Book 1–8*. Cambridge, MA and London: Loeb Classical Library, Harvard University Press.

Miller, J. I. (1969). *The Spice Trade of the Roman Empire 29 BC to AD 641*. Oxford: Clarendon Press.

Miller, N. F. and Gleason, K. L. (1994). *The Archaeology of Garden and Field*. Philadelphia: University of Pennsylvania Press.

Moisan, M. (1990a). *Lexique du vocabulaire botanique d'Hippocrate*. Québec: Laboratoire de recherches hippocratiques: Université Laval.

Moisan, M. (1990b). 'Les plantes narcotiques dans le corpus hippocratique'. In P. Potter, G. Maloney and J. Desautels (Eds), *La maladie et les maladies dans la collection hippocratique. Actes du VIe colloque international hippocratique (Québec, du 28 septembre au 3 octobre 1987)* (pp. 381–392). Québec: Les éditions du Sphinx.

Moraux, P. (1973). *Der Aristotelismus bei den Griechen, von Andronikos bis Alexander von Aphrodisias*. Berlin: De Gruyter.

Morton, A. G. (1981). *A History of Botanical Science: An Account of the Development of Botany from Ancient Times to the Present Day*. London: Academic Press.

Morton, A. G. (1986). 'Pliny on plants: his place in the history of botany'. In R. French and F. Greenaway (Eds), *Science in the Early Roman Empire: Pliny the Elder, His Sources and Influence* (pp. 86–97). London: Croom Helm.

Murphy, T. M. (2004). *Pliny the Elder's Natural History: The Empire in the Encyclopedia*. Oxford: Oxford University Press.

Murr, J. F. (1890). *Die Pflanzenwelt in der griechischen Mythologie*. Innsbruck: Wagnerische Universitäts-Buchhandlung.

Nachmanson, E. (1918). *Erotiani vocum Hippocraticarum collectio cum fragmentis*. Upspala: Appelbergs boktryckerei-aktiebolag.

Negbi, M. (1995). 'Male and female in Theophrastus' botanical works'. *Journal of the History of Biology*, 28, 317–332.

Nicole, J. (1891). *Les scolies genevoises de l'Iliade*. 2 vols. Paris: Librairie Hachette.

Nielsen, I. (2001). 'The gardens of the Hellenistic palaces'. In I. Nielsen (Ed.), *The Royal Palace Institution in the First Millennium BC: Regional Development and Cultural Interchange between East and West* (pp. 165–188). Athens: Danish Institute.

Nutton, V. (1991). 'Style and context in the *Method of Healing*'. In F. Kudlien and R. J. Durling (Eds), *Galen's Method of Healing* (pp. 1–25). Leiden: Brill.

Nutton, V. (1992). 'Healers in the medical market place: towards a social history of Graeco-Roman medicine'. In A. Wear (Ed.), *Medicine in Society: Historical Essays* (pp. 15–58). Cambridge: Cambridge University Press.

Nutton, V. (1993). 'Roman medicine: tradition, confrontation, assimilation'. In W. Haase and H. Temporini (Eds), *Aufstieg und Niedergang der römischen Welt. Teil II. Principat. Band 37.1.* (pp. 49–78). Berlin and New York: De Gruyter.

Nutton, V. (2010). '*De virtute centaureae*: a neglected Methodist text?' In D. Langslow and B. Maire (Eds), *Body, Disease and Treatment in a Changing World: Latin Texts and Contexts in Ancient and Medieval Medicine. Proceedings of the Ninth International Conference 'Ancient Latin Medical Texts', Hulme Hall, University of Manchester, 5th–8th September 2007*. Lausanne: Éditions BHMS.

Nutton, V. (2013). *Ancient Medicine*. Second edition. London: Routledge.

Oldfather, C. H. (1933). *Diodorus. Library of History. Volume I. Books 1–2.34*. Cambridge, MA and London: Loeb Classical Library, Harvard University Press.

Olivieri, A. (1935). *Aetii Amideni Libri medicinales I–IV. edidit A. Olivieri. CMG VIII 1.* Leipzig and Berlin: Akademie-Verlag.

Olivieri, A. (1950). *Aetii Amideni libri medicinales V–VIII. Edidit A. Olivieri. CMG VIII 1*. Berlin: Akademie-Verlag.

Olson, S. D. (1991). 'Firewood and charcoal in Classical Athens'. *Hesperia*, 60(3), 411–420.

Olson, S. D. (2007). *Athenaeus. The Learned Banqueters. Volume I. Books 1–3.106e. Edited and Translated by S.D. Olson*. Cambridge, MA and London: Loeb Classical Library, Harvard University Press.

Opsomer-Halleux, C. (1989). *Index de la pharmacopée du Ier au Xe siecle*. 2 vols. Hildesheim: Olms-Weidmann.

Orengo, H. A., Palet, J. M., Ejarque, A., Miras, Y. and Riera, S. (2013). 'Pitch production during the Roman period: an intensive mountain industry for a globalised economy?' *Antiquity*, 87(337), 802–814.

Orofino, G. (1991). 'Dioskurides war gegen Pflanzenbilde'. *Die Waage*, 4, 144–149.

Ørsted, A. S. (1869). 'Et Bidrag til Tydning af den Oldtider under Navn af Silfion meget anvendte og høit skaltede, men senere forsvundne Krydenplante'. *Oversigt over det Kongelige Danske Videnskabernes Selskabs forhandlinger og dets medlemmers arbeider*, 21–25.

Oser-Grote, C. M. (2004). *Aristoteles und das Corpus Hippocraticum: Die Anatomie und Physiologie des Menschen*. Stuttgart: Steiner.

Page, D. L. (1941). *Select Papyri in Four Volumes. vol. 3. Literary Papyri. Poetry. Texts, Translations and Notes by D.L. Page*. London and Cambridge, MA: Harvard University Press.

Pagnoux, C., Celant, A., Coubray, S., Fiorentino, G. and Zech-Matterne, V. (2013). 'The introduction of Citrus to Italy, with reference to the identification problems of seed remains'. *Vegetation History and Archaeobotany*, 22, 421–438.

Panagiotou, S. (1975). 'Aristotle's *de anima*, 424b3'. *Symbolae Osloenses*, 50(1), 47–53.

Parejko, K. (2003). 'Pliny the Elder's silphium: first recorded species extinction'. *Conservation Biology*, 17(3), 925–927.

Patera, M. (1994). 'Les rites d'extraction des plantes dans l'Antiquité. Magie, botanique et religion. L'exemple de la Mandragore'. *Revue des Archéologues et Historiens d'Art de Louvain*, 27, 21–34.

Paton, W. R. (1916). *The Greek Anthology. Vol. I. With an English Translation by W.R. Paton*. Cambridge, MA and London: Loeb Classical Library, Harvard University Press.

Pease, A. S. (1927). 'The loves of the plants'. *Classical Philology*, 22(1), 94–98.

Pease, A. S. (1933). 'Notes on ancient grafting'. *Transactions and Proceedings of the American Philological Association*, 64, 66–76.

Pease, A. S. (1952). 'A sketch of the development of ancient botany'. *Phoenix*, 6(2), 44–51.

Peck, A. L. (1942). *Aristotle. Generation of Animals. With an English Translation by A.L. Peck*. Cambridge, MA and London: Loeb Classical Library, Harvard University Press.

Peck, A. L. (1965). *Aristotle. Historia animalium. Books I–III. With an English Translation by A.L. Peck*. Cambridge, MA and London: Loeb Classical Library, Harvard University Press.

Peck, A. L. (1970). *Aristotle. Historia animalium. Books IV-VI. With an English Translation by A.L. Peck*. Cambridge, MA and London: Loeb Classical Library, Harvard University Press.

Peck, A. L. and Forster, E. S. (1961). *Aristotle. Volume XII. Parts of Animals. With an English Translation by A.L. Peck. Movement of Animals. Progression of Animals. With an English Translation by E.S. Forster*. Revised edition. Cambridge, MA and London: Loeb Classical Library, Harvard University Press.

Phillips, O. (2002). 'The witches' Thessaly'. In P. Minecki and M. Meyer (Eds), *Magic and Ritual in the Ancient World* (pp. 378–385). Leiden: Brill.

Pollard, E. A. (2009). 'Pliny's *Natural History* and the Flavian *Templum Pacis*: botanical imperialism in first-century CE Rome'. *Journal of World History*, 20(3), 309–338.

Polome, E. C. (1993). 'Echanges de vocabulaire entre le monde Méditerranéen et l'Europe septentrionale: les noms de plantes'. In *Actes du colloque international 'Les phytonymes grecs et latins' tenu à Nice les 14, 15, et 16 mai 1992 à la Faculté des Lettres, Arts et Sciences humaines, Université de Nice-Sophia Antipolis* (pp. 255–267). Nice: Centre de recherches comparatives sur les langues de la Méditerranée ancienne.

Polunin, O. (1980). *Flowers of Greece and the Balkans: A Field Guide*. Oxford: Oxford University Press.

Popa, I. C. (2010). 'The lists of plant synonyms in *De materia medica* of Dioscorides'. *Global Journal of Science Frontier Research*, 10(3), 46–49.

Porter, B. N. (1993). 'Sacred trees, date palms, and the royal persona of Ashurnasirpal II'. *Journal of Near Eastern Studies*, 52(2), 129–139.

Potter, P. (2012). *Hippocrates. Volume X. Edited and Translated by P. Potter*. Cambridge, MA and London: Loeb Classical Library, Harvard University Press.

Powell, O. (2003). *Galen: On the Properties of Foodstuffs. Introduction, Translation and Commentary, with a Foreword by John Wilkins*. Cambridge: Cambridge University Press.

Preisendanz, C. and Henrichs, A. (1973). *Papyri graecae magicae. Die griechischen Zauberpapyri. Herausgegeben und übersetzt von Karl Preisendanz. Zweite, verbesserte*

Auflage. Mit Ergänzungen von Karl Preisendanz. Durchgesehen und herausgegeben von Albert Henrichs. 2 vols. Stuttgart: Teubner.

Rackham, H. (1950). *Pliny. Natural History. Books 17–19. With an English Translation by H. Rackham.* Cambridge, MA and London: Loeb Classical Library, Harvard University Press.

Rackham, H. (1968). *Pliny. Natural History. Books XII–XVI. With an English Translation by H. Rackham.* Revised edition. Cambridge, MA and London: Loeb Classical Library, Harvard University Press.

Rackham, O. (1996). 'Ecology and pseudo-ecology: the example of ancient Greece'. In G. Shipley and J. Salmon (Eds), *Human Landscapes in Classical Antiquity: Environment and Culture* (pp. 16–43). London and New York: Routledge.

Radice, B. (1969). *Pliny. Letters and Panegyricus. Volume I. Letters, books I–VII. With an English Translation by B. Radice*. Cambridge, MA and London: Loeb Classical Library, Harvard University Press.

Raeder, J. (1929). *Oribasii Collectionum medicarum reliquiae. Libri IX–XVI. Edidit I. Raeder CMG VI 1,2.* Leizpig and Berlin: Teubner.

Raschke, M. C. (1978). 'New studies in Roman commerce with the East'. In H. Temporini (Ed.), *Aufstieg une Niedergang der Römischen Welt. II Principat. Band 9.2.* (pp. 604–1361) Berlin and New York: De Gruyter.

Raudnitz, J. M. (1843). *Materia medica Hippocratis: Dissertatio inauguralis*. Dresden: BG Teubner for the author.

Raven, F. (Ed.). (2000). *Plants and Plant Lore in Ancient Greece. With accompanying Essays by Alice Lindsell et al.* Oxford: Leopard's Head.

Raven, J. E. (1990). 'Plants and plant lore in Ancient Greece'. *Annales Musei Goulandris*, 8, 129–180. Reprinted in Raven, F. 2000: 5–39.

Reed, H. S. (1942). *A Short History of the Plant Sciences*. New York: Ronald Press.

Regenbogen, O. (1937). Eine Polemik Theophrasts gegen Aristoteles. *Hermes*, *72*(4), 469–475.

Regenbogen, O. (1940). 'Theophrastos'. In G. Wissowa, W. Kroll and K. Mittelhaus (Eds), *Paulys Real-Encyclopädie der classischen Altertumswissenschaft. Neue Bearbeitung. Supplmentband VII. Adobogiona bis Τριακαδιεῖς. Mit Nachträgen* (pp. 1354–1562). Stuttgart: J. B. Metzler.

Repici, L. (2006). 'Medici e botanica popolare'. In A. Marcone (Ed.), *Medicina e società nel mondo antico: atti del convegno di Udine (4–5 ottobre 2005)*. Firenze: Le Monnier università.

Reveal, J. M. (1996). 'What's in a name: identifying plants in prelinnaean botanical literature'. In B. K. Holland (Ed.), *Prospecting for Drugs in Ancient and Medieval European Texts: A Scientific Approach* (pp. 57–90). Australia: Harwood Academic Publishers.

Richlin, A. (1992). *The Garden of Priapus: Sexuality and Aggression in Roman Humor.* Oxford: Oxford University Press.

Riddle, J. M. (1981). 'Pseudo-Dioscorides' *Ex herbis feminis* and early medieval botany'. *Journal of the History of Biology*, 14, 43–81.

Riddle, J. M. (1985). *Dioscorides on Pharmacy and Medicine*. Austin: University of Texas Press.

Riggsby, A. M. (2007). 'Guides to the wor(l)d'. In J. König and T. Whitmarsh (Eds), *Ordering Knowledge in the Roman Empire* (pp. 88–107). Cambridge: Cambridge University Press.

Rival, M. (1991). *La charpenterie navale romaine: matériaux, méthodes, moyens*. Paris: Editions du CNRS.

Robinson, E. S. G. (1927). *Catalogue of the Greek Coins of Cyrenaica*. London: The British Museum.

Rodgers, R. H. (1975). *Palladii Rutilii Tauri Aemiliani Opus agriculturae; De veterinaria medicina; De insitione*. Leipzig: Teubner.

Rodgers, R. H. (2008a). 'Cassianus Bassus'. In P. T. Keyser and G. L. Irby-Massie (Eds), *The Encyclopedia of Ancient Natural Scientists: The Greek Tradition and its Many Heirs* (p. 207). London: Routledge.

Rodgers, R. H. (2008b). 'Florentinus'. In P. T. Keyser and G. L. Irby-Massie (Eds), *The Encyclopedia of Ancient Natural Scientists: The Greek Tradition and its Many Heirs* (p. 331). London: Routledge.

Rodgers, R. H. (2008c). 'L. Iunius Moderatus Columella of Gades'. In P. T. Keyser and G. L. Irby-Massie (Eds), *The Encyclopedia of Ancient Natural Scientists: The Greek Tradition and its Many Heirs* (pp. 456–457). London: Routldge.

Rodgers, R. H. (2008d). 'Palladius Rutilius Taurus Aemilianus'. In P. T. Keyser and G. L. Irby-Massie (Eds), *The Encyclopedia of Ancient Natural Scientists: The Greek Tradition and its Many Heirs* (pp. 35–36). London: Routledge.

Rodgers, R. H. (2008e). 'Sex. Quintilii, Condianus and Valerius Maximus, of Alexandria Troas'. In P. T. Keyser and G. L. Irby-Massie (Eds), *The Encyclopedia of Ancient Natural Scientists: The Greek Tradition and its Many Heirs* (p. 716). London: Routlegde.

Rolfe, J. C. (1940). *Ammianus Marcellinus. History. Volume II. Books 20–26. With an English Translation by J.C. Rolfe*. Cambridge, MA and London: Loeb Classical Library, Harvard University Press.

Romm, J. S. (1989). 'Aristotle's elephant and the myth of Alexander's scientific patronage'. *American Journal of Philology*, 110(4), 566–575.

Roques, D. (1984). 'Synésios de Cyrène et le silphion de Cyrénaique'. *Revue des Etudes Grecques*, 97, 380–399.

Roques, D. (1993). 'Médecine et botanique: le silphion dans l'oeuvre d'Oribase'. *Revue des Etudes Grecques*, 106, 380–399.

Rose, V. (1886). *Aristotelis qui ferebantur librorum fragmenta*. Leipzig: Teubner.

Ross, D. O. (1980). '*Non sua poma*: Varro, Virgil, and grafting'. *Illinois Classical Studies*, 5, 63–71.

Runnels, C. N. (1995). 'Environmental degradation in ancient Greece'. *Scientific American*, 272(3), 96–99.

Ryholt, K. (2013). 'The illustrated herbal from Tebtunis: new fragments and archaeological context'. *Zeitschrift für Papyrologie und Epigraphik*, 187, 233–238.

Sacconi, A. (1969). 'La mirra nella preparazione degli unguenti profumati a Cnosso'. *Athenaeum*, 47, 281–289.

Sadori, L., Allevato, E., Bosi, G., Caneva, G., Castiglioni, E., Celant, A., Di Pasquale, G., Giardini, G., Malzanti, M., Rinaldi, R., Rottoli, M. and Susanna, F. (2009). 'The introduction and diffusion of peach in ancient Italy'. In J.-P. Morel and A. M. Mercuri (Eds), *Plants and Culture: Seeds of the Cultural Heritage of Europe. Edipuglia* (pp. 45–61). Bari: Centro Europeo per I Beni Culturali.

Sallares, R. (1991). *The Ecology of the Ancient Greek World*. London: Duckworth.

Samama, E. (2006). '*Thaumatopoioi pharmakopôlai*: la singulière image des préparateurs et vendeurs de remèdes dans les textes grecs'. In F. Collard and E. Samama (Eds),

Pharmacopoles et apothicaires. Les 'pharmaciens' de l'Antiquité au Grand Siècle (pp. 7–27). Paris: l'Harmattan.

Sands, R. (2013). 'A history of human interactions with forests'. In R. Sands (Ed.), *Forestry in a Global Context* (pp. 1–36). Canterbury, NZ: CABI.

Sangin, M. A. F. (1936). *Codices Rossici. Catalogus Codicum Astrologorum Graecorum 12*. Brussels: Lamertin.

Sargeant, J. (1920). *The Trees, Shrubs, and Plants of Virgil*. Oxford: Blackwell.

Scarborough, J. (1984). 'Early Byzantine pharmacology'. *Dumbarton Oaks Papers*, 38, 213–232.

Scarborough, J. (1986). 'Pharmacy in Pliny's *Natural History*: some observations on substances and sources'. In R. French and F. Greenaway (Eds), *Science in the Early Roman Empire: Pliny the Elder, His Sources and Influence* (pp. 59–85). London: Croom Helm.

Scarborough, J. (1987). 'Adaptation of folk medicines in the formal materia medica of classical antiquity'. In J. Scarborough (Ed.), *Folklore and Folk-Medicines: Symposium Presented at the AIHP Annual Meeting Held in S. Francisco, CA in March 1986* (pp. 21–32). Madison: American Institute of the History of Pharmacy.

Scarborough, J. (1991). 'The pharmacology of sacred plants, herbs, and roots'. In C. A. Faraone and D. Obbink (Eds), *Magika Hiera: Ancient Greek Magic and Religion* (pp. 138–174). New York: Oxford University Press.

Scarborough, J. (2006a). 'Drugs and drug lore in the time of Theophrastus: folklore, magic, notany, philosophy and the rootcutter'. *Acta Classica*, 49, 1–29.

Scarborough, J. (2006b). 'More on Dioscorides' Etruscan herbs'. *Etrsucan News*, 6, 1 and 9.

Scarborough, J. (2008a). 'Aetios of Amida'. In P. T. Keyser and G. L. Irby-Massie (Eds), *The Encyclopedia of Ancient Natural Scientists: The Greek Tradition and its Many Heirs* (pp. 38–39). London: Routledge.

Scarborough, J. (2008b). 'Antonius Musa'. In P. T. Keyser and G. L. Irby-Massie (Eds), *The Encyclopedia of Ancient Natural Scientists: The Greek Tradition and its Many Heirs* (p. 101). London: Routledge.

Scarborough, J. (2008c). 'Areios of Tarsos, Laecanius'. In P. T. Keyser and G. L. Irby-Massie (Eds), *The Encyclopedia of Ancient Natural Scientists: The Greek Tradition and its Many Heirs* (pp. 128–129). London: Routledge.

Scarborough, J. (2008d). 'Asklēpiadēes of Bithunia'. In P. T. Keyser and G. L. Irby-Massie (Eds), *The Encyclopedia of Ancient Natural Scientists: The Greek Tradition and its Many Heirs* (pp. 170–171). London: Routledge.

Scarborough, J. (2008e). 'Caelius Aurelianus of Sicca'. In P. T. Keyser and G. L. Irby-Massie (Eds), *The Encyclopedia of Ancient Natural Scientists: The Greek Tradition and its Many Heirs* (pp. 201–202). London: Routledge.

Scarborough, J. (2008f). 'Diodotos (Pharm.)'. In P. T. Keyser and G. L. Irby-Massie (Eds), *The Encyclopedia of Ancient Natural Scientists: The Greek Tradition and its Many Heirs* (pp. 250–251). London: Routledge.

Scarborough, J. (2008g). 'Erasistratos of Ioulis on Keōs'. In P. T. Keyser and G. L. Irby-Massie (Eds), *The Encyclopedia of Ancient Natural Scientists: The Greek Tradition and its Many Heirs* (pp. 294–296). London: Routledge.

Scarborough, J. (2008h). 'Iulius Bassus'. In P. T. Keyser and G. L. Irby-Massie (Eds), *The Encyclopedia of Ancient Natural Scientists: The Greek Tradition and its Many Heirs* (p. 451). London: Routledge.

Scarborough, J. (2008i). 'Nikēratos (of Athens?)'. In P. T. Keyser and G. L. Irby-Massie (Eds), *The Encyclopedia of Ancient Natural Scientists: The Greek Tradition and its Many Heirs* (pp. 575–576). London: Routledge.

Scarborough, J. (2008j). 'Petrōnios Musa'. In P. T. Keyser and G. L. Irby-Massie (Eds), *The Encyclopedia of Ancient Natural Scientists: The Greek Tradition and its Many Heirs* (p. 639). London: Routledge.

Scarborough, J. (2008k). 'Scribonius Largus'. In P. T. Keyser and G. L. Irby-Massie (Eds), *The Encyclopedia of Ancient Natural Scientists: The Greek Tradition and its Many Heirs* (pp. 728–729). London: Routledge.

Scarborough, J. (2008l). 'Sextius Niger'. In P. T. Keyser and G. L. Irby-Massie (Eds), *The Encyclopedia of Ancient Natural Scientists: The Greek Tradition and its Many Heirs* (pp. 738–739). London: Routledge.

Scarborough, J. (2008m). 'Themisōn of Laodikeia (Syria)'. In P. T. Keyser and G. L. Irby-Massie (Eds), *The Encyclopedia of Ancient Natural Scientists: The Greek Tradition and its Many Heirs* (pp. 782–783). London: Routledge.

Scarborough, J. (2008n). 'Thessalos of Tralleis'. In P. T. Keyser and G. L. Irby-Massie (Eds), *The Encyclopedia of Ancient Natural Scientists: The Greek Tradition and its Many Heirs* (pp. 804–805). London: Routledge.

Scarborough, J. (2010). 'Attalus III of Pergamon: research toxicologist'. In L. Cilliers (Ed.), *Asklepios: Studies on Ancient Medicine* (pp. 138–156). Bloemfontien: Classical Association of South Africa.

Scarborough, J. (2011). 'Ancient medicinal use of Aristolochia: birthwort's tradition and toxicity'. *Pharmacy in History*, 53(1), 3–21.

Scarborough, J. (2012a). 'Crateuas'. In S. Hornblower, A. Spawforth and E. Eidinow (Eds), *The Oxford Classical Dictionary*. Fourth edition. Oxford: Oxford University Press.

Scarborough, J. (2012b). 'Nicander'. In S. Hornblower, A. Spawforth and E. Eidinow (Eds), *The Oxford Classical Dictionary*. Fourth edition. Oxford: Oxford University Press.

Scarborough, J. and Nutton, V. (1982). 'The preface of Dioscorides' *Materia Medica*: introduction, translation, commentary'. *Transactions and Studies of the College of Physicians of Philadelphia*, 4(3), 187–227.

Schneider, H. G. (1818). *Theophrasti Eresii quae supersunt opera. Tomus tertius. Annotationes*. Leipzig: Vogelli.

Schneider, O. (1856). *Nicandrea. Theriaca et Alexipharma recensuit et emendauit, fragmenta collegit, commentationes addidit Otto Schneider*. Leipzig: Teubner.

Scott, A. (1991). 'Ps-Thessalus of Tralles and Galen's *De Methodo Medendi*'. *Sudhoffs Archiv*, 75(1), 106–110.

Segura Munguía, S. and Torres Ripa, J. (2009). *Historia de las plantas en el mundo antiguo*. Bilbao: Universidad de Deusto.

Senn, G. (1929). 'Hat Aristoteles eine selbstandige Schrift unber Pflanzen verfasst?' *Philologus*, 85, 113–139.

Senn, G. (1942). 'Oak galls in the *Historia Plantarum* of Theophrastus'. *Transactions of the Royal Society of Edinburgh*, 60, 343–354.

Settis, S. (2008). *La Villa di Livia: le pareti ingannevoli*. Milan: Electa.

Shackleton Bailey, D. R. (1993). *Martial. Epigrams. Edited and Translated by D.R. Shackleton Bailey*. Cambridge, MA and London: Loeb Classical Library, Harvard University Press.

Sharples, R. W. (1992). 'Theophrastus: *On Fish*'. In W. W. Fortenbaugh and D. Gutas (Eds), *Theophrastus: His Psychological, Doxographical, and Scientific Writings* (pp. 347–385). New Brunswick: Transaction Publishers.

Sharples, R. W. (1995). *Theophrastus of Eresus: Sources for his Life, Writings, Thought and Influence. Commentary. Volume 5. Sources on Biology (Human Physiology, Living Creatures, Botany: Texts 328–435)*. Leiden: E. J. Brill.

Sharples, R. W. and Minter, D. W. (1983). 'Theophrastus on fungi: inaccurate citations in Athenaeus'. *Journal of Hellenic Studies*, 103, 154–156.

Shelmerdine, C. W. (1985). *The Perfume Industry of the Mycenaean Pylos*. Göteborg: P. Åströms Förlag.

Siber, U. G. (1699). *Moly, Hermetis herbam praecipvis nostrae vrbis doctoribvs magistris senatoribvs patronis et cvratoribus literarvm*. Sneeberga.

Sibthorp, J., Lindley, J. and Smith, J. E. (1806–1840). *Flora Græca, sive Plantarum rariorum historia, quas in provinciis aut insulis Græciæ legit, investigavit, et depingi curavit J. Sibthorp. Hic ille insertae sunt pauculae species, quas vir idem clarissimus, Graeciam versus navigans in itinere praesertim apud Italiam et Siciliam, invenerit. Characteres omnium, descriptiones et synonyma elaboravit Jacobus Edvardus Smith.* 10 vols. London: Taylor.

Simpson, M. G. (2010). *Plant Sytematics*. Second edition. Amsterdam: Elsevier.

Singer, C. (1927). 'The herbal in antiquity and its transmission to later ages'. *Journal of Hellenic Studies*, 47(1), 1–52.

Skoda, F. (1993). 'Principes de formation du lexique botanique en grec ancien: illustrations et hypothèses'. In *Actes du colloque international 'Les phytonymes grecs et latins' tenu à Nice les 14, 15, et 16 mai 1992 à la Faculté des Lettres, Arts et Sciences humaines, Université de Nice-Sophia Antipolis* (pp. 271–283). Nice: Centre de recherches comparatives sur les langues de la Méditerranée ancienne.

Smith, A. E. and Secoy, D. M. (1975). 'Forerunners of pesticides in classical Greece and Rome'. *Journal of Agricultural and Food Chemistry*, 23(6), 1050–1055.

Smith, W. D. (1990). *Hippocrates. Pseudepigraphic Writings. Letters – Embassy – Speech from the Altar – Decree. Edited and Translated with an Introduction by W. D. Smith.* Leiden: Brill.

Sollenberger, M. G. (1985). 'Diogenes Laertius 5.36–57: the *Vita Theophrasti*'. In W. W. Fortenbaugh, P. Huby and A. A. Long (Eds), *Theophrastus of Eresus: On His Life and Works* (pp. 1–62). New Brunswick: Transaction Books.

Sollenberger, M. G. (1988). 'Identifications of titles of botanical works of Theophrastus'. In W. W. Fortenbaugh and R. W. Sharples (Eds), *Theophrastean Studies: On Natural Sciences, Physics and Metaphysics, Ethics, Religion and Rhetoric* (pp. 14–24). New Brunswick: Transaction Books.

Sollenberger, M. G. (2008). 'Theophrastos of Eresos'. In P. T. Keyser and G. L. Irby-Massie (Eds), *The Encyclopedia of Ancient Natural Scientists: The Greek Tradition and its Many Heirs* (pp. 798–801). London: Routledge.

Somfai, A. (2008). 'Isidorus (Isidore) of Hispalis (Seville)'. In P. T. Keyser and G. L. Irby-Massie (Eds), *The Encyclopedia of Ancient Natural Scientists: The Greek Tradition and its Many Heirs* (p. 445). London: Routledge.

Sommerstein, A. H. (2009). *Aeschylus. Fragments. Edited and Translated by A. H. Sommerstein.* Cambridge, MA and London: Loeb Classical Library, Harvard University Press.

Spencer, D. (2010). *Roman Landscape: Culture and Identity*. Cambridge: Cambridge University Press.

Sprague, R. K. (1999). 'Plants as Aristotelian substances'. *Illinois Classical Studies*, 16(2), 221–229.

Sprengel, K. P. J. (1807). *Historia rei herbariae.* Volume 1. Amsterdam: Tabern.

Sprengel, K. P. J. (1817–1818). *Geschichte der Botanik.* Altenburh: Brockhaus.

Sprengel, K. P. J. (1822). *Theophrast's Naturgeschichte der Gewächse. Uebersetzt und erläutert von K. Sprengel.* Altona: J. F. Hammerich.

Stackhouse, J. (1811). *Illustrationes Theophrasti: in usum botanicorum praecipue peregrinantium.* Oxford: E Typographeo Clarendoniano.

Stackhouse, J. (1813–1814). *Theophrasti Eresii De historia plantarum libri decem: graece. Cum syllabo generum et specierum, glossario, et notis.* 2 vols. Oxford: J. Parker.

Stadler, H. (1898). 'Lateinische Pflanzennamen im Dioskorides'. *Archiv für lateinische Lexicographie*, 10, 83–115.

Stannard, J. (1965). 'Pliny and Roman botany'. *Isis*, 56, 420–425.

Stannard, J. (1969). 'The herbal as a medical document'. *Bulletin of the History of Medicine*, 43(3), 212.

Stannard, J. (1971). 'Byzantine botanical lexicography'. *Episteme*, 5(3), 168–187.

Stearn, W. T. (1973). *Botanical Latin: History, Grammar, Syntax, Terminology and Vocabulary.* Second edition. Newton Abbot: David and Charles.

Stearn, W. T. (1976). 'From Theophrastus and Dioscorides to Sibthorp and Smith: the background and origin of the Flora Graeca'. *Biological Journal of the Linnean Society*, 8, 285–298.

Stok, F. (2008a). 'A. Cornelius Celsus'. In P. T. Keyser and G. L. Irby-Massie (Eds), *The Encyclopedia of Ancient Natural Scientists: The Greek Tradition and its Many Heirs* (pp. 217–219). London: Routledge.

Stok, F. (2008b). 'Hērakleidēs of Taras (Med.)'. In P. T. Keyser and G. L. Irby-Massie (Eds), *The Encyclopedia of Ancient Natural Scientists: The Greek Tradition and its Many Heirs* (pp. 370–371). London: Routledge.

Stok, F. (2008c). 'Marcellus of Bordeaux, "Empiricus"'. In P. T. Keyser and G. L. Irby-Massie (Eds), *The Encyclopedia of Ancient Natural Scientists: The Greek Tradition and its Many Heirs* (pp. 527–530). London: Routledge.

Strömberg, R. (1937). *Theophrastea: Studien zur botanischen Begriffsbildung.* Göteborg: Wettergren and Kerbers.

Stubbings, F. H. (1946). 'Xerxes and the plane-tree'. *Greece and Rome*, 15(44), 63–67.

Tait, W. J. (1977). *Papyri from Tebtunis in Egyptian and in Greek.* London: Egypt Exploration Society.

Taub, L. (2004). *Ancient Meteorology.* London: Routledge.

Thanos, C. A. (2005a). 'The geography of Theophrastus' life and of his botanical writings'. In A. J. Karamanos and C. A. Thanos (Eds), *Biodiversity and Natural Heritage in the Aegean. Proceedings of the conference 'Theophrastus 2000' (Eressos-Sigri, Lesbos, July 6–8, 2000)* (pp. 113–131). Athens: Frangoudis.

Thanos, C. A. (2005b). 'Theophrastus on Oaks'. *Botanika Chronika*, 18, 29–36.

Thibodeau, P. (2008a). 'Attalos III of Pergamon, Philomētōr'. In P. T. Keyser and G. L. Irby-Massie (Eds), *The Encyclopedia of Ancient Natural Scientists: The Greek Tradition and its Many Heirs* (pp. 179–180). London: Routledge.

Thibodeau, P. (2008b). 'Dionusios of Utica, Cassius'. In P. T. Keyser and G. L. Irby-Massie (Eds), *The Encyclopedia of Ancient Natural Scientists: The Greek Tradition and its Many Heirs* (p. 265). London: Routledge.

Thibodeau, P. (2008c). 'Hostilius Saserna and son'. In P. T. Keyser and G. L. Irby-Massie (Eds), *The Encyclopedia of Ancient Natural Scientists: The Greek Tradition and its Many Heirs* (pp. 422–423). London: Routledge.

Thibodeau, P. (2008d). 'M. Porcius Cato of Tusculum'. In P. T. Keyser and G. L. Irby-Massie (Eds), *The Encyclopedia of Ancient Natural Scientists: The Greek Tradition and its Many Heirs* (pp. 686–688). London: Routledge.

Thibodeau, P. (2008e). 'M. Terentius Varro of Reate'. In P. T. Keyser and G. L. Irby-Massie (Eds), *The Encyclopedia of Ancient Natural Scientists: The Greek Tradition and its Many Heirs* (pp. 774–778). London: Routledge.

Thibodeau, P. (2008f). 'C. Tremellius Scrofa'. In P. T. Keyser and G. L. Irby-Massie (Eds), *The Encyclopedia of Ancient Natural Scientists: The Greek Tradition and its Many Heirs* (pp. 815–816). London: Routledge.

Thibodeau, P. (2008g). 'P. Vergilius Maro of Mantua'. In P. T. Keyser and G. L. Irby-Massie (Eds), *The Encyclopedia of Ancient Natural Scientists: The Greek Tradition and its Many Heirs* (pp. 824–827). London: Routledge.

Thibodeau, P. (2011). *Playing the Farmer: Representations of Rural Life in Vergil's Georgics*. Berkeley: University of California Press.

Thomas, R. (1992). *Literacy and Orality in Ancient Greece*. Cambridge: Cambridge University Press.

Thommen, L. (2012). *An Environmental History of Ancient Greece and Rome. Translated by Philip Hill.* Cambridge: Cambridge University Press.

Thompson, D. B. (1937). 'The garden of Hephaistos'. *Hesperia*, 6(3), 396–425.

Thompson, D. B. and Griswold, R. E. (1963). *Garden Lore of Ancient Athens*. Princeton: American School of Classical Studies at Athens.

Thompson, D. J. (1984). 'Hellenistic science: its application in peace and war: agriculture'. In F. W. Walbank and A. E. Astin (Eds), *The Cambridge Ancient History. Volume 7. Part 1. The Hellenistic World*. Second edition (pp. 364–370). Cambridge: Cambridge University Press.

Thomson, M. H. (1933). 'Catalogue des manuscrits grecs de Paris contenant des traités anonymes de botanique'. *Revue des Etudes Grecques*, 46, 334–348.

Thomson, M. H. (1955). *Textes grecs inédits relatifs aux plantes*. Paris: Les Belles Lettres.

Todd, F. A. (1943). 'Some Cucurbitaceae in Latin literature'. *Classical Quarterly*, 36(3–4), 101–111.

Tortzen, C. G. (1991). 'Male and female in peripatetic botany'. *Classica et mediaevalia*, 42, 81–110.

Totelin, L. M. V. (2009). *Hippocratic Recipes: Oral and Written Transmission of Pharmacological Knowledge in Fifth-and Fourth-Century Greece*. Leiden: Brill.

Totelin, L. M. V. (2012a). 'Botanizing rulers and their herbal subjects: plants and political power in Greek and Roman literature'. *Phoenix*, 66(1–2), 122–144.

Totelin, L. M. V. (2012b). 'And to end on a poetic note: Galen's authorial strategies in the pharmacological books'. *Studies in History and Philosophy of Science Part A*, 43(2), 307–315.

Totelin, L. M. V. (2015). 'When foods become remedies in ancient Greece: the curious case of garlic and other substances'. *Journal of Ethnopharmacology*, 167, 30–37.

Totelin, L. M. V. (forthcoming a). '*Pharmakopōlai*: a re-evaluation of the sources'. In W. Harris (Ed.), *Popular Medicine in Classical Antiquity*. Cambridge: Cambridge University Press.

Totelin, L. M. V. (forthcoming b). 'The birds and the bees: animal and plant generation in antiquity'. In R. Flemming, N. Hopwood, and L. Kassell (Eds), *Cultures of Reproduction from Antiquity to the Present Day*. Cambridge: Cambridge University Press.

Tournefort, J. P. de (1694). *Elemens de botanique ou méthode pour connoître les plantes*. Paris: Imprimerie royale.

Touwaide, A. (1991). 'Nicandre: de la science à la poésie: contribution à l'exégèse de la poésie médicale grecque'. *Aevum*, 65(1), 65–101.

Touwaide, A. (1993). 'Phytonymie et pharmacologie. Contribution à la sémantique du lexique phytonymique grec ancien à travers le traité de matière médicale de Dioscoride'. In *Actes du colloque international 'Les phytonymes grecs et latins' tenu à Nice les 14, 15, et 16 mai 1992 à la Faculté des Lettres, Arts et Sciences humaines, Université de Nice-Sophia Antipolis* (pp. 291–305). Nice: Centre de recherches comparatives sur les langues de la Méditerranée ancienne.

Touwaide, A. (1997). 'Bibliographie historique de la botanique: les identifications de plantes médicinales citées dans les traités anciens après l'adoption du système de Linné (1707–1778)'. *Centre Jean Palerne, Lettre d'informations*, 30, 2–22.

Touwaide, A. (1999a). 'La botanique entre science et culture au 1er siècle de notre ère'. In G. Wöhrle (Ed.), *Geschichte der Mathematik und der Naturwissenschaften in der Antike. Band 1. Biologie* (pp. 219–252). Stuttgart: Steiner.

Touwaide, A. (1999b). 'Lexica medico-botanica byzantina. Prolégomènes à une étude'. In Lois C. Pérez Castro, F. R. Adrados and L. A. de Cuenca (Eds), *Miscelánea léxica en memoria de Conchita Serrano* (pp. 211–228). Madrid: Consejo superior de Investigaciones Cientificas.

Touwaide, A. (2007). 'Art and sciences: private gardens and botany in the early Roman empire'. In M. Conan (Ed.), *Botanical Progress, Horticultural Innovation, and Cultural Changes* (pp. 37–49). Washington, DC: Dumbarton Oaks.

Trépanier, S. (2008). 'Empedoklēs of Akragas'. In P. T. Keyser and G. L. Irby-Massie (Eds), *The Encyclopedia of Ancient Natural Scientists: The Greek Tradition and its Many Heirs* (pp. 283–284). London: Routledge.

Triller, D. W. (1716). *Moly Homericum detectum, cum reliquis ad fabulam Circaeam pertinentibus*. Leipzig.

Uden, J. (2010). 'The vanishing gardens of Priapus'. *Harvard Studies in Classical Philology*, 105, 189–219.

van der Eijk, P. J. (1997). 'Galen's use of the concept of "qualified experience" in his dietetic and pharmacological works'. In A. Debru (Ed.), *Galen on Pharmacology: Philosophy, History, and Medicine: Proceedings of the Vth International Galen Colloquium, Lille, 16–18 March 1995* (pp. 35–57). Leiden: Brill.

van der Eijk, P. J. (2000). *Diocles of Carystus: A Collection of the Fragments with Translation and Commentary*. Volume I. Leiden: Brill.

van der Eijk, P. J. (2001). *Diocles of Carystus: A Collection of the Fragments with Translation and Commentary*. Volume 2. *Commentary*. Leiden: Brill.

van der Veen, M., Livarda, A. and Hill, A. (2008). 'New plant foods in Roman Britain—dispersal and social access'. *Environmental Archaeology*, 13(1), 11–36.

Vercoutre, M. A. T. (1913). 'Le silphium des anciens est bien un palmier (*Ladoicea sechellarum De Labillardière*)'. *Revue générale de botanique*, 25, 31–37.

Viano, C. (1992). 'Théophraste, Ménestor de Sybaris et la συμμετρία de la chaleur'. *Revue des Etudes Grecques*, 105, 584–592.

Vogt, S. (2008). 'Dāmokratēs, Seruilius'. In P. T. Keyser and G. L. Irby-Massie (Eds),

The Encylcopedia of Ancient Natural Scientists: The Greek Tradition and its Many Heirs (p. 226). London: Routledge.

Vogt, S. (2009). 'Drugs and pharmacology'. In R. J. Hankinson (Ed.), *The Cambridge Companion to Galen* (pp. 304–322). Cambridge: Cambridge University Press.

Voigts, L. E. (1978). 'The significance of the name Apuleius to the *Herbarium Apulei*'. *Bulletin of the History of Medicine*, 52, 214–227.

von Stackelberg, K. T. (2009). *The Roman Garden: Space, Sense, and Society*. London: Routledge.

von Staden, H. (1975). 'Experiment and experience in Hellenistic medicine'. *Bulletin of the Institute of Classical Studies*, 22, 178–199.

von Staden, H. (1996). 'Liminal perils: early Roman receptions of Greek medicine'. In F. J. Ragep, S. P. Ragep and S. Livesey (Eds), *Tradition, Transmission, Transformation: Proceedings of two Conferences on Pre-Modern Science Held at the University of Oklahoma* (pp. 369–418). Leiden: Brill.

Wallraff, M., Scardino, C., Mecella, L., Guignard, C. and Adler, W. (2012). *Iulius Africanus: Cesti: The Extant Fragments*. Berlin: De Gruyter.

Wedel, G. W. (1688). *Propempticon inaugurale de tetragono Hippocratis*. Jena: Krebs.

Wedel, G. W. (1713). *Propempticon inaugurale de moly Homeri*. Jena: Krebs.

Wedel, G. W. (1714). *Propempticon inaugurale de holoconitide Hippocratis*. Jena: Krebs.

Wehrli, F. (1969). *Die Schule des Aristoteles. Texte und Kommentar. Herausgegeben von F. Wehrli. Hft. 9. Phainias von Eresos, Chamaileon, Praxiphanes*. Second edition. Basel and Stuttgart: Schwabe.

Weinstock, S. (1953). *Codices Britannici. Catalogus Codicum Astrologorum Graecorum 9.2*. Brussels: Lamertin.

Wellmann, M. (1889). 'Sextius Niger, eine Quellenuntersuchung zu Dioscorides'. *Hermes*, 24(4), 530–569.

Wellmann, M. (1898a). 'Das älteste Kräuterbuch der Griechen'. In *Festgabe für Franz Susemihl. Zur Geschichte griechischer Wissenschaft und Dichtung* (pp. 1–31). Leipzig: Teubner.

Wellmann, M. (1898b). 'Die Pflanzennamen des Dioskurides'. *Hermes*, 33(3), 360–422.

Wellmann, M. (1903). 'Dioskurides 12'. In G. Wissowa (Ed.), *Paulys Real-Encylopädie der classischen Altertumswissenschaft. Neue Bearbeitung. Neunter Halband. Demogenes-Donatianus* (pp. 1131–1142). Stuttgart: J. B. Metzler.

Wellmann, M. (1906). *Pedanii Dioscuridi Anazarbei De materia medica libri quinque. Edidit M. Wellmann. Volumen II quo continentur libri III et IV*. Berlin: Weidmann.

Wellmann, M. (1907). *Pedanii Dioscuridis Anazarbei De materia medica libri quinque. Edidit M. Wellmann. Volumen I quo continentur libri I et II*. Berlin: Weidmann.

Wellmann, M. (1914). *Pedanii Dioscuridis Anazarbei De materia medica libri quinque. Edidit M. Wellmann. Volumen III quo continentur liber V, Crateuae et Sixtii Nigri fragmenta, Dioscuridis libri de simplicibus*. Berlin: Weidmann.

Wellmann, M. (1924). 'Beiträge zur Quellenanalyse des Älteren Plinius'. *Hermes*, 59(2), 129–156.

Wendland, P. (1901). *Alexandri in librum de sensu commentarium*. Berlin: Reimer.

Wertime, T. A. (1983). 'The furnace versus the goat: the pyrotechnologic industries and Mediterranean deforestation in antiquity'. *Journal of Field Archaeology*, 10(4), 445–452.

White, K. D. (1970). *Roman Farming*. London: Thames & Hudson.

Wilberding, J. (2008). 'Porphyry and Plotinus on the seed'. *Phronesis: A Journal for Ancient Philosophy*, 53(4–5), 4–5.

Wilberding, J. (2011). *Porphyry. To Gaurus on how Embryos are Ensouled and on what is in our Power*. London: Bristol Classical Press.

Wilkins, J. (2004). 'The social and intellectual context of *Regimen II*'. In P. J. van der Eijk (Ed.), *Hippocrates in Context: Papers Read at the XIth International Hippocrates Colloquium, University of Newcastle Upon Tyne, 27–31 August 2002* (pp. 121–133). Leiden: Brill.

Wimmer, F. (1838). *Phytologiae Aristotelicae fragmenta*. Vratislaviae: Typis Grassii, Barthii et socii.

Winter, W. (1951). 'Two Greek names for the truffle'. *American Journal of Philology*, 72, 63–64.

Wöhrle, G. (1985). *Threophrasts Methode in seinen botanischen Schriften*. Amsterdam: Grüner.

Wöhrle, G. (1988). 'The structure and function of Theophrastus' treatise *De odoribus*'. In W. W. Fortenbaugh and R. W. Sharples (Eds), *Theophrastean Studies: On Natural Sciences, Physics and Metaphysics, Ethics, Religion and Rhetoric* New Brunswick: Transaction Books.

Wöhrle, G. (1995). 'Aristoteles als Botaniker'. In W. Kullmann and S. Föllinger (Eds), *Aristotelische Biologie. Intentionen, Methoden, Ergebnisse. Akten des Symposions über Aristoteles' Biologie vom 24. bis 28* (pp. 387–396). Stuttgart: Steiner.

Zhmud, L. (2008a). 'Hippōn of Krotōn'. In P. T. Keyser and G. L. Irby-Massie (Eds), *The Encyclopedia of Ancient Natural Scientists: The Greek Tradition and its Many Heirs* (p. 421). London: Routledge.

Zhmud, L. (2008b). 'Menestōr of Subaris'. In P. T. Keyser and G. L. Irby-Massie (Eds), *The Encyclopedia of Ancient Natural Scientists: The Greek Tradition and its Many Heirs* (pp. 547–548). London: Routledge.

Zohary, D., Hopf, M. and Weiss, E. (2013). *Domestication of Plants in the Old World*. Fourth edition. Oxford: Oxford University Press.

Zotter, H. (1980). *Antike Medizin: die medizinische Sammelhandschrift Cod. Vindobonensis 93 in lateinischer und deutscher Sprache*. Graz: Akademische Druck- und Verlagsanstalt.

Zucker, A. (2008a). 'Iouba II of Mauretania, C. Iulius'. In P. T. Keyser and G. L. Irby-Massie (Eds), *The Encyclopedia of Ancient Natural Scientists: The Greek Tradition and its Many Heirs* (pp. 441–442). London: Routledge.

Zucker, A. (2008b). 'Nikolaos of Damaskos'. In P. T. Keyser and G. L. Irby-Massie (Eds), *The Encyclopedia of Ancient Natural Scientists: The Greek Tradition and its Many Heirs*. London: Routledge.

Zucker, A. (2008c). 'Philōn of Alexandria'. In P. T. Keyser and G. L. Irby-Massie (Eds), *The Encyclopedia of Ancient Natural Scientists: The Greek Tradition and its Many Heirs* (p. 653). London: Routledge.

PASSAGES CITED

Inscriptions and papyri

INDEX OF PLANTS

Page numbers in *italics* denote tables, those in **bold** denote figures.

Each index entry includes an English plant name, followed by a possible Linnaean identification. Often, that identification is of a genus rather than a species (e.g. *Acanthus* sp.), either because we discuss several species in the book, or because it is not possible to determine a precise species. Plants named by analogy with another are indexed together with their analogues. For instance, 'sea oak' is indexed under 'oak'. We have also indexed the Greek or Latin plants names that were left untranslated in this book; these names are italicised and their language (Greek and/or Latin) is indicated in brackets. Both within this book and here, we talk of 'lotos', 'moly' and 'silphium', three very important plants that are almost impossible to identify, without using italics. References to large groups of plants (or plant-like organisms), such as 'cereals', 'ferns' and 'fungi' are to be found in the General Index. We have also provided a key to the index of plants, where we have listed Linnaean names in alphabetic order.

Key to the index of plants: scientific names

GENERAL INDEX

Page numbers in *italics* denote tables, those in **bold** denote figures.